I0048641

MEDICINAL PLANT ALKALOIDS

An Introduction for Pharmacy Students

STEPHEN K. SIM

Associate Professor of Pharmacognosy
Faculty of Pharmacy
University of Toronto

Second Edition

UNIVERSITY OF TORONTO PRESS

Second Edition

© Copyright 1965 by

The University of Toronto Press

Reprinted in 2018

Reprinted 1966

First Edition, 1964

ISBN 978-0-8020-1355-2 (paper)

The use of portions of the text of the United States Pharmacopeia Seventeenth Revision, official September 1, 1965, is by permission received from the Board of Trustees of the United States Pharmacopeial Convention. The said Board is not responsible for any inaccuracies of the text thus used.

Permission to use portions of the text of the National Formulary, Twelfth Edition, official September 1, 1965, has been granted by the American Pharmaceutical Association. The American Pharmaceutical Association is not responsible for any inaccuracy of quotation, or for false implications that may arise by reason of the separation of excerpts from the original context.

Printed in Canada

PREFACE TO THE SECOND EDITION

Among the medicinally useful constituents of plants, the alkaloids constitute an important group. In discussing the medicinal plant alkaloids as part of a phytochemically oriented course in pharmacognosy conducted by the author for students in Pharmacy, he was faced with the difficulty of recommending to the students a suitable reference book which would discuss, at an introductory level, the basic phytochemical aspects of the medicinally important alkaloids, particularly their properties in relation to their separation and isolation from the plant sources. Discussions of these aspects are of course to be found in the more advanced and comprehensive books on alkaloids or the special monographs on particular groups of alkaloids, as well as in various scientific journals. But these more comprehensive books and journals are generally beyond easy comprehension by undergraduate students who have not had a more elementary introduction to these subjects, and are, at any rate, rather voluminous.

The accounts and discussions given here attempt to summarize (a) the plant sources and fundamental properties of the medicinally important alkaloids, (b) the principles and certain useful methods for their separation and isolation from plant materials, and (c) some aspects of their biosynthesis. They are intended as an introduction to these aspects of the study of alkaloids, as a supplement to the relevant sections on the alkaloids in standard text-books of pharmaceutical subjects, and to provide a collateral background for related laboratory works. It is hoped that these introductory accounts may also serve to prepare the student for proceeding to independent and effective use of the more comprehensive literature on these and other alkaloids.

Those alkaloids which are rather rarely used today for medicinal purposes have been omitted. On the other hand, a few alkaloids which are not medicinally useful but which occur in the same plant source as the medicinally useful ones or which serve to illustrate some phytochemical points are included.

Accounts of biosynthesis are confined to only four groups, namely, the Nicotiana alkaloids, the tropane alkaloids, the ergot alkaloids, and the opium alkaloids. These have been selected for discussion on the basis of their having relatively better-established experimental evidence, and on the basis of their suitability as examples for illustrating this aspect of the medicinal plant alkaloids.

Only brief indications of the principal pharmacological actions and uses are given for the purpose of relating these alkaloids to broad categories of their medicinal uses. The principal preparations in the United States Pharmacopeia, the National Formulary, the British Pharmacopoeia, and the British Pharmaceutical Codex, and a few pharmaceutical specialties are listed simply by name as examples of medicinal agents containing the alkaloids and some of their derivatives. Examples of pharmaceutical specialties are given by the names prevailing in Canada and/or the United States. Some of these preparations are not necessarily available in both countries. No exhaustive coverage is attempted for the plant sources, or for the chemical aspects, or for the medicinal preparations.

A preliminary edition was privately printed in 1964 for the use of Pharmacy students at the University of Toronto through the much appreciated assistance of the Book Department of the University of Toronto Press. In this second edition, Chapter 7 is new. Chapters 2, 4 and 8 have been considerably revised and slightly expanded. The section on the biosynthesis of Nicotiana alkaloids in Chapter 3 has been entirely re-written. Editorial modifications and minor revisions have been made in all other chapters, and the literature references are now given at the end of each chapter.

For permission to quote copyright materials, I am grateful to the Academic Press, J. B. Lippincott Co., the Ronald Press Co., and Springer-Verlag. My thanks are also due to the University of Toronto Press for invaluable assistance and co-operation. I am grateful to my wife for her assistance in secretarial work in the preparation of the draft and the index.

S. K. S.

Toronto,

July, 1965

CONTENTS

CHAPTER 1

INTRODUCTION

Thousands of plant species have been investigated with respect to their alkaloidal constituents. Hundreds of alkaloids have been chemically characterized, and many of these have been chemically synthesized. Most of the plant alkaloids possess in varying degrees pharmacological activities of one kind or another. However, a large number of the pharmacologically active alkaloids are not medicinally useful either because their activity is too feeble or because their toxicity is too marked. These considerations have in practice restricted the number of the medicinally useful alkaloids to a few scores or so. Those alkaloids which are pharmacologically or thereapeutically useful constitute a very important group of medicinal agents.

These introductory discussions will be largely confined to those alkaloids which are of current pharmacological or medicinal importance. A few alkaloids which are of little medicinal importance but which occur in the same plant species as the medicinally important ones and which affect the isolation and extraction of the latter are also discussed. The basic chemical aspects and properties of these alkaloids are discussed in relation to the plant sources from which they may be isolated.

It may be of interest to note that, while chemical synthesis has very largely replaced the plant source in the supply of a number of alkaloids for medicinal uses (e.g., ephedrine, quinine), in a number of other cases the plant sources continue to be the important raw material sources after chemical synthesis has been achieved, largely for economic reasons (e.g., reserpine, ergonovine, morphine).

It is also to be expected that from time to time fuller chemical and pharmacological studies of new as well as old alkaloids may result in the introduction of more alkaloids to medicinal use while the medicinal importance of some of the currently useful alkaloids may decrease as a result of the availability of better agents, natural or synthetic, for similar therapeutic purposes.

CHAPTER 2

ALKALOIDS - General Considerations

2.1 General Characteristics and Properties of Alkaloids

Since the isolation of morphine by Sertürner in 1817, there have been several hundred alkaloids isolated from various plants and the chemical constitutions for many of these have been elucidated. Total chemical synthesis has also been achieved for scores of these alkaloids, although in a number of these cases the biological source continues to be the most economical way of producing them. Thousands of different plant species have been investigated within the past two decades for alkaloidal constituents (18). However, only a relatively small number among all the known alkaloids are currently of importance from the therapeutic or pharmacological point of view.

The alkaloids have been traditionally treated as a group, and yet there has been no completely satisfactory definition of an alkaloid. This is so, because the term alkaloid is not a chemical designation but rather a name traditionally and conventionally accepted for a group of nitrogen-containing, basic substances from plants with rather widely different chemical constitutions. Nevertheless, there are several common features and attributes, which, to a greater or lesser degree, are associated with or possessed by the compounds commonly known as alkaloids. Among all the known alkaloids, cases of exceptions to one or more of these features can be cited. These features are as follows: (1) Alkaloids are more or less complex compounds produced by plants. (2) They contain nitrogen in the molecule. (3) Many, probably most, alkaloids are derived, in their biosynthesis, at least partly from various amino acids as their direct precursors. (4) They are basic (alkaline) in reaction. (5) They are usually soluble in a number of immiscible organic solvents, but rather insoluble in water. (6) Many of them are precipitated by certain reagents. (7) A number of them give more or less characteristic colour reaction with certain reagents. (8) Most of them are susceptible to destruction by heat; a number of them undergo decomposition or degradation by exposure to air and/or light. (9) A great number of alkaloids show rather pronounced pharmacological actions on various organs and tissues of animals and humans.

In the following discussion of these properties, it will become obvious that these properties are of considerable practical importance in the course of extraction and isolation of alkaloids from plant materials and in the considerations of pharmaceutical production and formulation involving alkaloids.

1. Plant Source

Alkaloids occur in many different species in numerous genera and families of vascular plants as well as in certain species of fungi. It has been estimated that some fifteen per cent or more of all vascular plants contain alkaloids. A number of amines produced by animals (e.g., epinephrine) possess physical and chemical properties rather similar to those of alkaloids. By tradition and convention, these animal amines are generally not considered as alkaloids. The occurrence of alkaloids in different plant organs and tissues and their relationship to aspects of the physiology of the plants will be discussed in a later section.

2. Nitrogen in the Molecule

Alkaloids contain one or more nitrogen atoms in the molecule. A

- 4 -

large number of alkaloids contain at least one nitrogen atom in a heterocyclic ring. But in a number of alkaloidal amines (e.g., ephedrine, mescaline, colchicine) which are generally considered as alkaloids, the nitrogen in the molecule is not in the ring.

3. Amino Acids as Biosynthetic Precursors

Certain alpha-amino acids have been experimentally shown to serve as precursors in the biosynthesis of many alkaloids, not only with respect to the nitrogen-containing portion of the alkaloid molecule, but in some cases also with respect to the non-nitrogenous moiety as well (15). Among the amino acids which have been either experimentally established or postulated on the basis of theoretical chemical considerations as metabolic precursors of various alkaloids, the most common ones are: phenylalanine, tyrosine, lysine, ornithine, histidine, tryptophan, and anthranilic acid. The precursor role of some of these will be discussed in later chapters in conjunction with discussions of the biosynthesis of certain groups of alkaloids. Some alkaloids (e.g., steroidal alkaloids) are, in their biogenesis, more directly derived from terpenoid or other precursor compounds of carbohydrate metabolism.

4. Alkalinity

The alkaloids are basic (alkaline) in reaction, due to the presence of the nitrogen in the molecule. The term "alkaloid" means alkali-like. Therefore they form salts with various acids. Most of the alkaloid bases are crystalline solids; a few are liquids (e.g., nicotine, pilocarpine). The alkaloid salts are crystalline, and the microscopic examination of their crystalline structure sometimes serves as an aid to their identification.

The alkaloid salts in solution release the free alkaloid bases when the solution is made alkaline, most commonly with ammonia, sodium carbonate or calcium hydroxide. The relative solubilities of the free alkaloid bases and of the alkaloid salts will be discussed below. These relative solubili ties and the interconversion of alkaloid bases and their corresponding salt constitute an important principle on which a variety of methods of extraction and isolation of alkaloids are based.

All alkaloids do not have the same degree of alkalinity. Apart fro the influence exerted on the electronic disposition of the nitrogen in the alkaloid molecule by side chains and various substitutions, one important factor contributing to the different degrees of alkalinity among different alkaloids is whether a given alkaloid contains primary, secondary, tertiary or quaternary nitrogen atom or atoms. Such differences in the degree of al· kalinity arising from the various structural characters are reflected in the different pK_a values for the different alkaloids. The weaker bases (those alkaloids with low pK_a values) would require a more acidic medium to form salts with the acid than would the strongly basic alkaloids (those with high pK_a values). Therefore, in a medium at a weakly acidic pH, some strongly basic alkaloids may be converted to their salt form by reacting with the acid present, while the alkaloids which are weaker bases (with lower pK_a) may still be in their free-base form. Such a situation is sometimes utilized in the separation of a particular alkaloid or a group of alkaloids with closely similar pK_a values from other alkaloids which have very much higher or very much lower pK_a values.

5. <u>Solubility</u>

Most <u>alkaloid bases</u> themselves are practically water-insoluble.

They are usually fairly soluble in a number of organic solvents such as chloroform, ether, and the lower alcohols. There are, of course, exceptions. For instance, morphine and colchicine (unlike most of other alkaloids) are practically insoluble in ether. Also, many alkaloids which contain phenolic hydroxyl group in the molecule are soluble in aqueous solutions of caustic alkali (e.g., morphine, cephaeline).

On the other hand, the alkaloid salts are generally soluble in water and in alcohol, and mostly nearly insoluble in the immiscible organic solvents.

A few alkaloid bases are rather soluble in water, but these are exceptions rather than the rule. Some examples of water-soluble alkaloids are: ephedrine, colchicine, ergonovine, and the clavine-type of ergot alkaloids (agroclavine, etc.). It should be noted, however, that these water-soluble alkaloids are still relatively more soluble in a number of organic solvents (such as chloroform or ether) than they are in water, and their salts are, again, nearly insoluble in the immiscible organic solvents.

In Table 1 are listed the approximate solubility data for a few examples of alkaloids and their salts. The contrast in the relative solubility between the alkaloid (base) atropine and its salt atropine sulphate typifies the majority of alkaloids and their salts, while ephedrine and ergonovine represent exceptions. In another respect, morphine (the alkaloid base) also represents an exception in that it is practically insoluble in ether and also rather sparingly soluble in chloroform; most alkaloids are very much more soluble in these two solvents.

Table 1. Relative Solubility of Some
Alkaloids and Their Salts

Alkaloid or its salt	ml. of solvent to dissolve 1 gm. of alkaloid or alkaloid salt at 25°C.			
	Water	Alcohol	Ether	Chloroform
Atropine	460	2	25	1
Atropine Sulphate	0.5	5	insoluble	insoluble
Morphine	3400	300	5000	1525
Morphine Sulphate	15.5	565	insoluble	insoluble
Ephedrine	35	soluble	soluble	soluble
Ergonovine	soluble	soluble	soluble	soluble
Colchicine	22	soluble	157	soluble

6. Precipitation by Certain Reagents

Many alkaloids in quantities as small as 10 - 100 μg. in solution
may form precipitate or turbidity with certain reagents (12). Most alkaloids are also precipitated by tannins. With special tecniques, some alkaloids in quantities smaller than 1 μg. may be detected in this way. On this
basis, these alkaloid-precipitating reagents are sometimes used in testing
the presence or absence of alkaloids in crude extracts of plant materials,
and in testing whether a particular step in an extraction procedure has exhausted the alkaloidal contents. A negative response (i.e., no precipitate
or turbidity) can be taken to mean absence of alkaloids, but a positive test
may or may not be due to the presence of alkaloids in the solution or extract
being tested, as some non-alkaloidal substances may also give such a posi-

tive response to these reagents.

Under controlled conditions of mixing and slow evaporation, a drop of an alkaloid solution reacting with a drop of an appropriate alkaloid-precipitating reagent on a microscope slide forms micro-crystals with rather characteristic shapes and manner of aggregation for a particular alkaloid reacting with a particular reagent. This is sometimes used as an aid to identification of the alkaloid (1, 3, 7).

Among the reagents most commonly used for testing alkaloids by precipitation or micro-crystal formation are: Mayer's (with several quantitative variations of composition), Wagner's (with quantitative variations of composition), Dragendorff's, Kraut's (modification of Dragendorff's), Marme's, Hager's, Scheibler's, gold chloride or bromide (5% aqueous solution), reineckate salt solution.

<u>Mayer's Reagent</u>:

Mercuric chloride	1.36 gm.
Potassium iodide	5.00 gm.
Water to make	100 ml.

<u>Wagner's Reagent</u>:

Iodine	1.3 gm.
Potassium iodide	2.0 gm.
Water to make	100 ml.

<u>Kraut's Reagent</u> (modification of <u>Dragendorff's</u>):

Bismuth nitrate	8.0 gm.
Nitric acid	20.0 ml.
Potassium iodide	27.2 gm.
Water to make	100 ml.

<u>Marme's Reagent</u>:

Cadmium iodide	10 gm.
Potassium iodide	20 gm.
Water to make	100 ml.

Scheibler's Reagent:

Sodium tungstate	20 gm.
Disodium phosphate	70 gm.
Water to make	100 ml.
Acidify with nitric acid.	

Hager's Reagent:

Saturated solution of picric acid.

Reineckate Salt Solution:

Dissolve 1 gm. of ammonium reineckate $NH_4 \left[Cr(NH_3)_2(SCN)_4 \right].H_2O$ and 0.3 gm. of hydroxylamine hydrochloride in 100 ml. of ethanol, filter, and store in refrigerator.

7. Colour Reaction with Certain Reagents

Certain groups of alkaloids give more or less characteristic colours with certain reagents. In some cases, under standardized conditions, the intensity of the colour so formed is in linear proportion to the concentration of the alkaloids in solution, and may be used in quantitative determination of those groups of alkaloids. The blue colour formed by the ergot alkaloids with the van Urk reagent (Ehrlich reagent, p-dimethylaminobenzaldehyde in 65% sulphuric acid) and the Vitali colour reaction of the Belladonna alkaloids (with fuming nitric acid and alcoholic potassium hydroxide solution) are such examples.

Apart from these rather specific colour reagents, there are a number of reagents which more generally elicit colour responses from a fairly wide variety of alkaloids. Although such colour reactions are rather unspecific, when used with proper technique, these reagents are often very sensitive so that in many cases only micro-gram quantities of the alkaloids are required to give these colour responses (3).

Examples of such reagents are: (a) a 0.5% solution selenium dioxide;

(b) a 0.5% solution of ammonium molybdate; (c) a saturated aqueous solution of ammonium vanadate; (d) Marquis reagent (concentrated sulphuric acid containing one drop of 40% formaldehyde per ml.); (e) a solution of 1 gm. of ceric ammonium sulphate in 99 gm. of 85% phosphoric acid (especially useful for indole alkaloids).

A modified Dragendorff Reagent (14), which is widely used for spraying paper chromatograms of many alkaloids, has the following composition:

1.6% solution of bismuth subnitrate in 20% acetic acid	5 ml.
40% aqueous solution of potassium iodide	5 ml.
Glacial acetic acid	20 ml.
Water to make	100 ml.

8. Sensitivity to Heat and Light (Stability)

Many alkaloids undergo decomposition or degradation when allowed to stand at temperatures above 70° C. or so for long periods of time. This sensitivity to heat and/or light varies in degree among different alkaloids. Decomposition or degradation occurs much more easily when an alkaloid is in solution than when it is in dry form. During isolation procedures, storage of dry alkaloidal extracts in vacuum desiccator over a drying agent such as phosphorus pentoxide or calcium chloride is a common practice, together with precautions against excessive exposure to heat and light. In the course of extraction the removal of solvent from an alkaloidal extract or solution may be effectively carried out by distillation under reduced pressure or by evaporation by a rotary evaporator assembly at temperatures considerably lower than the boiling point of the solvent.

9. Pharmacological Activity

Most alkaloids exert some definite pharmacological action. In many cases, a small quantity of alkaloid brings about a rather pronounced pharmacological effect on various organs and tissues of animal and human bodies. Potency varies among different alkaloids. For instance, the therapeutic doses (for humans) of a few medicinally useful alkaloids are: atropine, 0.25 - 1 mg.; reserpine, 0.25 - 1 mg. (for antihypertensive action) and 2 - 4 mg. (for psychosis); morphine, 8 - 20 mg. (oral); papaverine, 120 - 250 mg. (oral); quinidine sulphate, 200 mg. (for cardiac arrhythmia); quinine sulphate, 1 gm. (as antimalarial).

10. Isomers

Many alkaloids contain one or more asymmetric carbon atoms in the molecule, and therefore show optical activity. In the majority of cases the (-)-isomer (laevorotatory) has considerably greater pharmacological activity than the (+)-isomer (dextrorotatory) of the same alkaloid.

The traditional designations l- and d- for the laevo- and dextrorotatory isomers respectively are to be distinguished from the designations L- and D- which refer, not to optical activity, but to the steric configuration in relation to a conventionally accepted reference compound.

A few examples may serve to illustrate the considerable differences in pharmacological activity among the different isomers of an alkaloid. For instance, the relative pressor activities of D(-)-ephedrine and D(+)-ephedrine are 36 and 11 respectively; that is, the D(-)-isomer is about three and a half times as active as the D(+)-isomer. Ergotamine (which is the (-)-form) possesses three to four times greater pharmacological action than ergotaminine (which is the (+)-isomer of ergotamine). Therefore, usually either the (-)-form or the racemic form (and not the (+)-form) is used for therapeutic or pharmacological purposes. Again, there are exceptions: The medicinally useful d-tubocurarine is the (+)-form. In some alkaloids, both the

(-)-form and the (+)-form are medicinally useful. Quinine (the (-)-form) and quinidine (which is the (+)-form of quinine) are such examples. Quinine is primarily used as an antimalarial, while quinidine is primarily used for its action in restoring cardiac arrhythmia to normal rhythm.

2.2 Extraction and Isolation of Alkaloids

In alkaloid-bearing plant materials, the alkaloids constitute only a small percentage, ranging (in terms of dry weight) from about 10% of morphine in opium, 5 - 8% of quinine in Cinchona bark, about 0.2% of hyoscyamine in Atropa belladonna leaf, to 0.1 - 0.2% of reserpine and less than 0.1% of rescinnamine in Rauwolfia serpentina root. In the extraction and isolation of the alkaloids from the plant materials, there is first the problem of separating the alkaloids from the bulk of the non-alkaloidal materials. Also, very few plants (if any) produce just one single alkaloid: in the great majority of alkaloid-bearing plants, several alkaloids, which are usually rather closely related to one another in chemical structures, occur in the same plant. Therefore, there is the further problem of separation of the individual alkaloids from a mixture of alkaloids.

Each individual alkaloid has its particular physical and chemical characteristics, such as its solubility in different solvents of varying polarity, its pK_a value, degree of ease of crystallization of its salt-forms from different solvents, etc. Therefore, the methods and techniques best suited to the isolation of each alkaloid ultimately differ from the methods which are most effective for the isolation of another alkaloid. However, the extraction and isolation of the total alkaloids (i.e., mixture of all or nearly all the alkaloids present in a given sample of plant material) reasonably free from inorganic and non-alkaloidal organic matter may be effected by certain general procedures which are fairly effectively applicable to the preliminary isolation of total alkaloids of a wide variety of chemical constitutions from a wide range of plant materials.

These general procedures are largely based on the alkaline nature of most alkaloids and the consequent ability to form salts with acids, and on the relative solubilities of the alkaloid bases and their salts in water and in various organic solvents.

Before any extraction is begun, the plant material needs to be reduced to a moderately coarse powder by a suitable means to facilitate effective contact of solvent with alkaloid-containing tissues and cells. In plant materials that are rich in fats or oils (especially seeds), these should be removed by preliminary extraction with a suitable fat-solvent (petroleum ether is often used) which would not extract the alkaloids in question. In some cases, this defatting process may consist of occasionally shaking the powdered plant material in the fat-solvent over a period of several hours and subsequent removal of the solvent. In other cases, the procedure may require extraction of the fats or oils from the plant material with the fat-solvent in a Soxhlet extractor for twenty-four hours or longer.

The procedures most commonly employed for the extraction of <u>total</u> <u>alkaloids</u> from plant materials may be grouped into two main types: (A) procedure using water-immiscible organic solvent for the initial extraction, and (B) procedure using aqueous or alcoholic medium for the initial extraction. In many cases, trace quantities of impurities present in solvents even of "analytical reagent" grade have to be removed by suitable chemical

or physical means before these solvents can be used for extraction of alkaloids (4, 11).

The alkaloids occur in the plant mostly as the salts. When the plant
material is placed in an alkaline medium or when an alkaloidal extract is
made alkaline, the alkaloid salts are converted to the corresponding alkaloid
bases. The relative solubility properties of the free alkaloid bases and of
their salts have been discussed earlier. Chloroform and ether are among the
most commonly used immiscible solvents for alkaloid extractions, although
for certain alkaloids some other organic solvents such as benzene, ethylene
chloride, carbon tetrachloride, etc., may be used with advantage for particular alkaloids. Chloroform extracts, with varying degrees of ease, all
alkaloids except the quaternary bases.

Procedure (A)

1. The powdered plant material is moistened with 10% ammonia or sodium carbonate solution, and then macerated for one-half hour or one hour
with the organic solvent to be used (say, chloroform) in a quantity just
sufficient to cover the powder.

2. The macerated powder and solvent are then transferred to a continuous extraction apparatus such as the Soxhlet extractor assembly. The
original container or flask is washed with suitable quantities of the same
solvent, and the washings added to the extractor unit. Continuous extraction is then carried out with a suitable quantity of the solvent over a
period of several hours. For small-scale extraction, in most cases with
10 - 20 gm. sample of plant material, a period of 3 - 6 hours is usually
sufficient. With larger batches (several Kg.) of plant material with the

use of giant-sized Soxhlet apparatus (10 - 20 litres capacity), the period required for this continuous extraction may be 24 - 48 hours.

3. The extract so obtained is filtered, and the filtrate is shaken, in a separatory funnel, with several portions of an aqueous acid solution. For this purpose, a 1% or 2% mineral acid solution (such as sulphuric acid) or a 1% or 2% organic acid (tartaric acid is commonly used) solution is generally employed. As pointed out earlier (p. 6), the pK_a range of the alkaloids being extracted needs to be considered in using a specific concentration of an acid for converting the alkaloid bases to the alkaloid salts.

If the pH of the aqueous acidic solution is suitable, the acid converts the free alkaloid bases to their salts. These salts are generally considerably more soluble in aqueous medium than in the organic solvent such as chloroform or ether (in many cases, insoluble in these solvents). Therefore, the volume of the acid solution to be used for each shaking need not be very large. For each shaking, 20 - 30 ml. of acid solution for the first shaking and 15 - 20 ml. for each of three or more subsequent shakings are usually sufficient for 100 ml. of the chloroform (or ether) extract. With proper technique, the shaking with the first portion of acid solution will usually extract 90% or more of the alkaloids (as salts) from the organic solvent into the aqueous acid. But in order to obtain as close to one hundred per cent extraction as possible, three or four additional portions of acid solution must be used. One of the alkaloid-precipitating reagents, such as Mayer's or Wagner's reagent, is sometimes used to test whether the last portion of the acid extract has exhausted the alkaloidal content of the chloroform or ether extract being extracted.

The shaking with each portion of the aqueous acid solution may be carried out by inverting back and forth the separatory funnel containing the mixture for a hundred times or more.

Insufficient mixing of the two phases would of course result in ineffective extraction of the alkaloids from one phase to the other. But excessively vigorous shaking of the mixture may cause formation of emulsion, especially with chloroform. If a small amount of emulsion is formed, usually as a layer between the organic solvent layer and the aqueous acid layer, it may often be effectively broken by centrifugation or by filtration of this emulsified inter-phase through dry filter paper.

A five to ten-minute period of standing may be necessary to allow complete separation of the two phases at the completion of shaking with each portion of the acid solution. Even when this is done, the aqueous acid extract would inevitably still contain some small amount of the organic solvent (chloroform or ether). For complete removal of the organic solvent from the aqueous acid solution, evaporation or distillation under reduced pressure may be used.

The combined aqueous acid extract so obtained may still contain some non-alkaloidal matter such as resins or colouring matter in addition to the alkaloid salts, but it is now sufficiently free from the bulk of non-alkaloidal substances to be used for certain colorimetric quantitative estimation of the alkaloids (e.g., in the case of the assay of Prepared Ergot) or for quantitative estimation by some other method.

4. Resins and certain colouring matter, if present, and other non-alkaloidal impurities may be removed from this aqueous acid solution by

shaking, <u>under acidic conditions</u>, in a manner similar to what has been described above, with portions of chloroform, and discarding the chloroform extract. Trace amounts of alkaloids may sometimes be lost by this treatment, but usually the amount so lost would be negligible.

5. The aqueous acidic (alkaloidal) solution in a separatory funnel is covered with a suitable quantity of the same organic solvent (chloroform) or of another solvent (say, ether), and 10% ammonia is slowly added to make the solution alkaline (or to a specific pH). Extraction is then carried out by shaking as before with several portions of the organic immiscible solvent. This may also be carried out in a continuous manner with the use of a liquid-liquid extraction apparatus. The alkalinity converts the alkaloid salts in the aqueous solution to the free base forms of the alkaloids and these latter, being usually (other than the exceptional cases) much more soluble in the organic solvent (say, chloroform or ether), pass into the organic solvent layer.

6. Even when a period of twenty to thirty minutes is allowed for the separation of the two phases in the separatory funnel after each extraction, the organic solvent extract so obtained is still not at all free from water. Removal of water from this extract may be achieved by stirring into it, or shaking it in a stoppered flask, with a quantity of anhydrous sodium sulphate (usually 1 - 2 gm. for 60 - 80 ml. of chloroform extract or 40 - 50 ml. of ether extract) for half an hour or longer. It is then filtered through dry filter paper to remove the sodium sulphate, and then again stirring the filtrate with another similar quantity of anhydrous sodium sulphate for a few minutes, and then placed in a vacuum desiccator over anhydrous calcium chloride over-night and again filtered.

7. The extract, now free of water, may be evaporated to dryness or the solvent may be removed by distillation under reduced pressure or by a rotary flash evaporator. This leaves a residue containing a mixture of the alkaloids from the original plant material (a total-alkaloids extract), reasonably free from non-alkaloidal inorganic and organic impurities.

This alkaloidal residue may be weighed, and a part or all of it may be dissolved in (taken up by) a suitable organic solvent, and used for paper or column chromatography by solvent systems that are suitable for separation of that particular group of alkaloids, or it may be taken up by a suitable quantity of aqueous acid solution and the resulting solution may then be used for certain quantitative determination procedures or for crystallization by suitable techniques.

Procedure (B)

1. The powdered plant material (previously defatted, if necessary) is shaken with alcohol on a mechanical shaker for several hours or over-night.

2. The alcoholic extract, containing alkaloids and non-alkaloidal organic matter, is concentrated to a suitable small volume, preferably under reduced pressure (which, however, may often cause foaming). The concentrated alcoholic extract is then acidified to pH 1 - 2, to convert the alkaloids to their salts. The alcohol is then removed by steam-distillation or distillation under reduced pressure, leaving the aqueous acidic solution (containing the alkaloid salts). This aqueous solution may usually contain water-insoluble organic matter (originally in solution in the alcohol) in suspension. In general 0.5 - 1 ml. of aqueous acid solution is used for each gm. of the plant material.

3. This aqueous extract is then allowed to stand in room temperature over-night, and then placed in a refrigerator for a further two to three-day period, to allow separation of resin material. The clear aqueous solution may then be decanted and filtered through filter paper or through charcoal, if necessary.

4. The aqueous extract from (2) may also be extracted, under acidic conditions, with portions of chloroform, and the chloroform extract containing resins, and fats and oils is discarded.

5. This aqueous extract may still contain water-soluble organic materials in addition to the alkaloid salts. It may now be further extracted as in steps 5, 6, and 7 of Procedure (A).

2.3 Distribution of Alkaloids in Plants

Occurrence in different organs of the plant

The occurrence of alkaloids in plants does not appear to be confined to certain specific organs. Alkaloids are found in various different organs (leaf, root, fruit, etc.) in various different plant species. It should be pointed out that in any one particular species usually only one or two organs, and not all organs, possess the function of alkaloid formation. For instance, the alkaloids of the tobacco plant (Nicotiana tabacum) are formed in the root and are translocated to the leaves where the alkaloids accumulate, while the seeds contain no alkaloid. In the opium poppy (Papaver somniferum) the alkaloids occur in the latex of the fruit, while the seeds are devoid of alkaloids. In the autumn crocus (Colchicum autumnale) the alkaloids occur in the seed as well as the corm. In the Cinchona species, the alkaloids accumulate in the bark.

In some cases there are fluctuations of alkaloid content in various organs of the plant during the growing season, or between day and night (8, 17). In other cases, especially perennials, localization of the alkaloids in one or two particular organs appears to become more marked with the increased age of the plant (13, 17).

Site of formation of alkaloids

When alkaloids are found to be present in particular organs or parts of a plant, it does not necessarily mean that the alkaloids are formed (synthesized) in those particular organs. The alkaloids in several Datura species and Nicotiana species are mostly formed in the roots, but are rapidly translocated to the leaves. That is, the site of formation of the alkaloids in these plants is the root, but the leaf, where the alkaloids accumulate, is the part to be used for the extraction and isolation of any appreciable quantities of the alkaloids.

The site of alkaloid formation in several Datura, Nicotiana, and Atropa species (all of the family Solanaceae) has been very well demonstrated by various investigators (5, 6, 10) using grafting experiments and other techniques. Thus, when Atropa belladonna scion (aerial parts without the root system) was grafted on Nicotiana tabacum stock (the part with the root system), the Nicotiana alkaloid (nicotine) was soon found to be distributed in the Atropa aerial parts.

The results of such experiments also show that the alkaloid hyoscyamine is largely produced in the root of Atropa and Datura species, but is produced in all parts of Datura innoxia. Scopolamine (hyoscine) is formed

- 21 -

mainly in the root of <u>Datura innoxia</u>, <u>Datura metel</u>, and of <u>Atropa belladonna</u>, but is formed in the aerial parts of <u>Datura meteloides</u>, and in all parts of <u>Datura ferox</u>.

<u>Physiological factors and alkaloids</u>

    <u>Isolated</u> roots of <u>Nicotiana</u> species, <u>Atropa belladonna</u>, and certain other species have been found to excrete alkaloids into the medium, but the roots of <u>intact plants</u> (of these same species) excrete very little alkaloids. In <u>Papaver somniferum</u> (opium poppy) the alkaloids are degraded in ripening fruits; hence, for obtaining opium, incisions are made on the <u>unripe</u> fruits. Alkaloid-forming functions in a particular plant may vary at different stages of development of the plant (13, 17). It has been mentioned earlier that the seeds of <u>Nicotiana</u> species and the seeds of <u>Papaver somniferum</u> contain no alkaloids. On germination, the alkaloids narcotine, codeine, morphine, and papaverine quickly appear.

    In those <u>Solanaceae</u> plants which produce both scopolamine (hyoscine) and hyoscyamine, the scopolamine/hyoscyamine ratio seems to be fairly constant in the mature plant in a given species. This ratio differs from one species to another. But in the young plants, especially in their roots, scopolamine often predominates even in those plants which contain predominantly hyoscyamine in the fully mature stage.

    By reciprocal grafts of various <u>Nicotiana</u> species, Jeffrey and Tso (9) showed that the summation (total amount) of alkaloids accumulated in the shoot and the root of plants grafted to a foreign scion (or to a foreign stock) was always lower than the amount of alkaloids found in the respective intact (non-grafted) plants.

A number of plant growth regulators which promote growth (such as gibberellins and others), when applied to various <u>Solanaceae</u> plants, mostly tend to decrease the percentage content of alkaloids, although certain species under certain conditions of treatment with these growth substances show increase in alkaloid contents (16). On the other hand, <u>Datura meteloides</u> when treated with certain "growth retardants" have shown some increase of alkaloidal content (2).

2.4 <u>References</u>

1. Association of Official Agricultural Chemists - Official Methods of Analysis of A.O.A.C., 9th ed., pp. 508-511, Washington, D.C., 1960.

2. Bennett, J. H., & Sciuchetti, L. A. - J. Pharm. Sci., 53:1254-1256 (1964).

3. Clarke, E. G. C. - J. Pharm. Pharmacol., 9:187-192 (1957).

4. Cromwell, B. T. - in K. Paech and M. V. Tracey (eds.): Modern Methods of Plant Analysis, Vol. IV, pp. 367-374, Springer, Berlin, 1955.

5. Dawson, R. F. - Adv. Enzymol., 8:203-251 (1948).

6. Dawson, R. F. - Am. Scientist, 48:321-340 (1960).

7. Farmilo, C. G., & Genest, K. - in C. P. Stewart and A. Stolman (eds.): Toxicology: Mechanism and Methods of Analysis, Vol. II, pp. 229-242, Academic Press, New York, 1961.

8. Hemberg, T., & Fluck, H. - Pharm. Acta Helv., 28:74 (1953).

9. Jeffrey, R. N., & Tso, T. C. - Plant Physiol., 39:480-483 (1964).

10. Leete, E. - in A. G. Avery, S. Satina, & J. Rietsema (eds.): Blakeslee: The Genus Datura, pp. 48-56. Ronald Press, New York, 1959.

11. Manske, R. H. F. - in R. H. F. Manske & H. L. Holmes (eds.): The Alkaloids, Vol. I, pp. 7-11, Academic Press, New York, 1950.

12. Martello, R., & Farnsworth, N. R. - Lloydia, 25:176-185 (1962).

13. Mothes, K. - in R. H. F. Manske (ed.) The Alkaloids, Vol. VI, pp. 1-29, Academic Press, New York, 1960.

14. Munier, R., & Macheboeuf, M. - Bull. Soc. Chim. Biol., 32:192 (1950).

15. Ramstad, E., & Agurell, S. - Ann. Rev. Plant Physiol., 15:143-168 (1964).

16. Sciuchetti, L. A. - J. Pharm. Sci., 50:981-998 (1961).

17. Sokolov, V. S. - Symp. Soc. Exp. Biol., No. 13, pp. 230-257, Cambridge University Press, London, 1959.

18. Willaman, J. J., & Schubert, B. G. - Alkaloid-Bearing Plants and Their Alkaloids, U. S. Dept. Agric., Technical Bull. No. 1234, Washington, 1961.

CHAPTER 3

ALKALOIDS OF THE PYRIDINE GROUP

3.1 Introduction

Among the alkaloids of pharmacological or therapeutic interest which
contain a pyridine ring or a reduced derivative of pyridine in the molecule
are the alkaloids of Nicotiana species and Areca species.

In the Nicotiana alkaloids, nicotine contains both a pyridine ring
and a pyrrolidine ring, while anabasine contains a pyridine ring and a piperidine ring. Arecoline and arecaidine, the two principal alkaloids of Areca
Catechu, contain neither a pyridine ring nor a piperidine ring, but a partially reduced pyridine derivative.

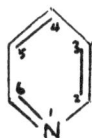

Pyridine

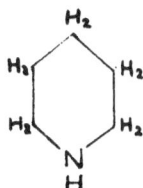

Piperidine

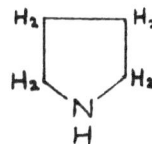

Pyrrolidine

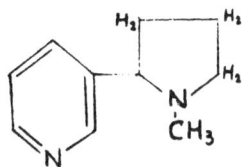

Nicotine

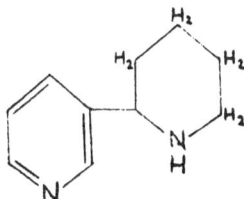

Anabasine

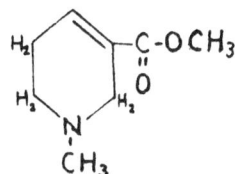

Arecoline

3.2 Alkaloids of Nicotiana Species

The leaves of the tobacco plant (Nicotiana tabacum L., Family Solanaceae) contain several alkaloids, of which nicotine (a liquid alkaloid) cons-

titutes about three-quarters of the total alkaloidal content. The leaf contains about 4 - 6% of nicotine. Among other alkaloids present are nor-nicotine and anabasine. Nicotiana rustica L. and certain other species of Nicotiana also contain these alkaloids. In Nicotiana glauca Graham, the ratio of nicotine to anabasine is about 1:4; that is, more or less the reverse of the ratio in Nicotiana tabacum. Nicotine and nor-nicotine also occur in the leaves of Duboisia hopwoodii F. v. Muell. (Fam. Solanaceae) to the extent of 1 - 2%. This is an Australian woody plant, sometimes known by the common name Pituri or Pitchery. The Nicotiana species are annual herbs.

The site of formation of nicotine in the Nicotiana plants has been shown to be in the root (3, 4, 9). When shoots (scion) of tomato plants (Lycopersicum esculentum Mill., Fam. Solanaceae) were grafted on to the root system (stock) of Nicotiana tabacum, the tomato shoots were found to contain the Nicotiana alkaloids. Conversely, very little nicotine was present in the Nicotiana leaves which were grown on tomato roots.

Properties of Nicotine

Pure nicotine is a colourless liquid alkaloid, b.p. 246°. It may be distilled unchanged in a current of steam. It is soluble in alcohol, ether, or petroleum ether. Its miscibility with water at the temperature range of 60° - 210° C. is limited, but it is miscible with water in all proportions below and above this temperature range. The salts of nicotine are readily water-soluble. Natural laevorotatory nicotine forms hydrochloride, sulphate, and tartrate salts which are dextrorotatory. The tartrate salt crystallizes from alcohol on addition of ether. Exposure of nicotine to ultraviolet light may convert it to nicotine oxide, nicotinic acid, and methylamine.

In the estimation of nicotine and nor-nicotine, the plant material
is dried at 60° C., treated with sodium hydroxide solution, and then the alkaloids are steam-distilled into a standardized acid solution and the uncom-
bined acid may then be titrated. Alternatively, the acid solution containing the alkaloids (nicotine and nor-nicotine) may be measured for absorbance
at 260 mµ.

Action and Uses of Nicotine

Nicotine's primary pharmacological action is on the sympathetic and
parasympathetic ganglia, consisting of a primary transient stimulation and a
more persistent depression leading to paralysis. It also has paralytic action on skeletal muscle, and other pharmacological effects. It is very toxic,
and very large doses may prove to be fatal in a few seconds (8).

The principal use of nicotine is as an insecticide, commonly used in
the form of a 40% solution of nicotine sulphate in various formulations.

Biosynthesis of Nicotiana Alkaloids

(a) The participation of nicotinic acid in the formation of the pyridine ring portion of nicotine was shown by the experiments of Dawson (4)
and Solt (16). In their experiments, H^3-labelled and C^{14}-labelled nicotinic
acid was administered to shoots of tobacco plants, and the isotopes were subsequently found to be incorporated in the pyridine ring of nicotine isolated
from the plants.

(b) When tryptophan-β-C^{14} was fed to the tobacco plant (2, 11, 13),
no C^{14} incorporation in the pyridine ring of the alkaloid was found. From
this it is concluded that the nicotinic acid (used by the tobacco plant in
the formation of the alkaloids) is not formed in these plants via the tryp-

tophan-kynurenine intermediates as it is in the case of certain fungi.

(c) Leete et al. (6, 14) fed Nicotiana glauca plants with acetate-2-C^{14} and glycerol-2-C^{14} (separately), subsequently isolated the anabasine formed by the plants, and degraded the alkaloid by permanganate oxidation to nicotinic acid (derived from the pyridine ring portion of the anabasine). Determination and study of the C^{14}-content of the individual carbons of the nicotinic acid so obtained (through a series of degradation processes) indicated that 1. acetate-2-C^{14} led to the same pattern of isotope-labelling in the nicotinic acid as would be resulted from succinic acid-2,3-C^{14}, 2. carbons 4, 5, and 6 of the nicotinic acid were derived from a three-carbon compound closely related to glycerol (they postulated that glyceraldehyde-3-phosphate might serve as this 3-carbon compound), and 3. the carbons 2, 3, and 7 of the nicotinic acid were derived from succinic acid or a metabolically closely related compound (postulated as aspartic acid).

From these experimental results, Leete (14) postulated that in these plants acetic acid (i.e., in its metabolically active form, acetyl-Co.A) through the Krebs cycle would give rise to succinic acid and then to oxaloacetic acid, which by transamination would then yield aspartic acid. Condensation between aspartic acid and glyceraldehyde-3-phosphate would form the heterocyclic ring and lead to quinolinic acid. Decarboxylation of quinolinic acid would then give nicotinic acid.

If this postulate is correct, then feeding the tobacco plant with aspartic acid-3-C^{14} should lead to the incorporation of the C^{14} isotope into carbon No. 3 of the pyridine ring portion of the alkaloid formed. Griffith et al. (7) fed aspartic acid-3-C^{14} to the tobacco plant and found that about

50% of the radioactivity of the pyridine ring of the isolated alkaloid was located at carbon No. 3 of the pyridine ring. Leete (14) explained this by postulating that the administered aspartic acid-3-C^{14} was metabolized through the Krebs cycle to the **symmetrical** succinic acid which would be equally labelled with C^{14} isotope at its No. 2 and No. 3 carbons. This succinic acid would again form (resynthesize) aspartic acid which would now be equally labelled with C^{14} at both No. 2 and No. 3 carbons; that is, each of carbons No. 2 and 3 of this resynthesized aspartic acid would now have only 50% of the C^{14} content derived from the original aspartate-3-C^{14}. Consequently carbon No. 3 of the pyridine ring which would be derived from carbon No. 3 of this resynthesized aspartic acid would have only about 50% of the radioactivity of the pyridine ring, the other 50% being presumably located at carbon No. 2 of the pyridine ring (derived from carbon No. 2 of the resynthesized aspartic acid). Leete's postulate is illustrated in Figure 1.

(d) The pyrrolidine ring of nicotine was shown to be derived from ornithine by the experiments of Byerrum et al. (5, 9, 13). Feeding ornithine-2-C^{14} to tobacco plants gave rise to nicotine which contained C^{14} label in the α-atoms (carbons 2 and 5) of its pyrrolidine ring.

(e) Leete (12) showed the incorporation of C^{14}-label into the nicotine formed by tobacco plants that had been fed putrescine-1,4-C^{14}. Leete et al. (15) further established that the nitrogen of the pyrrolidine ring of nicotine was derived from the δ-amino, and not the α-amino, of ornithine. By feeding ornithine-2-C^{14} and ornithine-δ-N^{15} (together) to root cultures of <u>Nicotiana tabacum</u> they (15) found the incorporation of the N^{15}-label into the pyrrolidine ring of the nicotine subsequently isolated, and no incorpo-

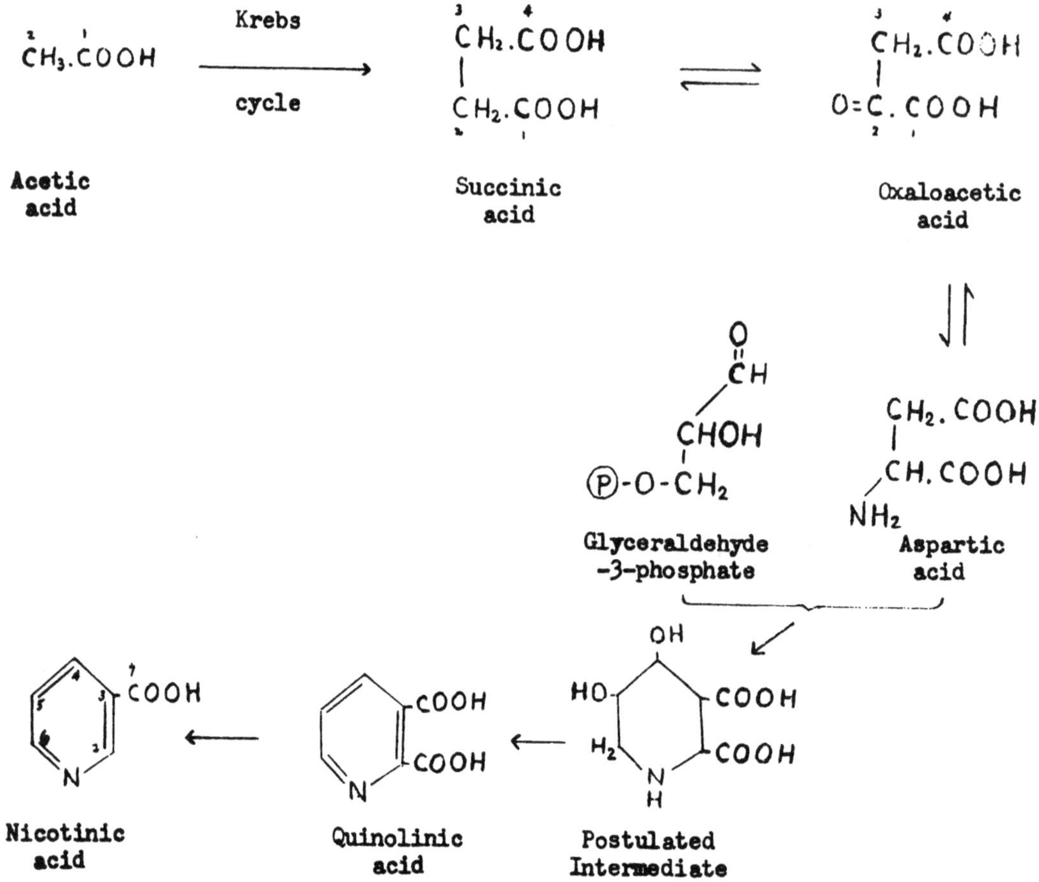

$\overset{2}{C}H_3.\overset{1}{C}OOH$ $\xrightarrow[\text{cycle}]{\text{Krebs}}$ $\overset{3}{C}H_2.\overset{4}{C}OOH$
$\overset{}{C}H_2.\overset{}{C}OOH$ $\rightleftharpoons$ $\overset{3}{C}H_2.\overset{4}{C}OOH$
$O=\overset{2}{C}.\overset{1}{C}OOH$

Acetic
acid

Succinic
acid

Oxaloacetic
acid

$\updownarrow$

$\overset{O}{\overset{\|}{C}}H$
$CHOH$
$\textcircled{P}\text{-O-}CH_2$

Glyceraldehyde
-3-phosphate

$CH_2.COOH$
$CH.COOH$
NH_2

Aspartic
acid

Nicotinic acid ← Quinolinic acid ← Postulated Intermediate (from Glyceraldehyde-3-phosphate and Aspartic acid)

Nicotinic
acid

Quinolinic
acid

Postulated
Intermediate

Figure 1. Leete's postulated biogenetic scheme for nicotinic
acid formation in Nicotiana plants

ration of the N^{15}-label into the pyridine ring of the alkaloid. When orni-
thine-α-N^{15} was fed (also together with ornithine-2-C^{14}) to such root cultures, no incorporation of the N^{15}-label into the nicotine occurred. In
these experiments the C^{14} label from the ornithine-2-C^{14} was incorporated
into carbons 2 and 5 of the pyrrolidine ring as was with the case of the intact plants quoted in (d) above.

These authors postulated two possible routes for the formation of the pyrrolidine ring from ornithine (schematically shown in Figure 2): One of these would involve α-transamination of ornithine to α-keto-δ-amino-valeric acid, and decarboxylation of the latter to 4-aminobutanal. Cyclization would then lead to Δ'-pyrroline, which would further lead to the pyrrolidine ring of nicotine. Quantitative considerations of their results necessitated their additional postulation that some δ-transamination of ornithine to glutamic semi-aldehyde probably also occurred, with re-synthesis of ornithine.[*] The alternative route postulated would involve decarboxylation of ornithine to putrescine and oxidative deamination of the latter (with loss of one of the two nitrogens) to 4-aminobutanal which would then cyclize to Δ'-pyrroline and finally lead to the formation of the pyrrolidine ring of nicotine.

(f) The pathway for the formation of the pyrrolidine ring of nicotine from carbon dioxide has also been investigated. By feeding separately acetate-1-C^{14}, acetate-2-C^{14}, propionate-2-C^{14}, and aspartate-3-C^{14} to Nicotiana rustica plants, in short period (1 - 2 hours) and long period (6 hours and 168 hours) experiments, and by studying the distribution patterns of the isotope in the nicotine and the intermediates formed (such as glutamate), Wu et al. (17) concluded from their experiments that the main pathway for the pyrrolidine ring from carbon dioxide is through glycolysis and the tricarboxylic acid cycle to glutamate, which is then converted to glutamic semi-aldehyde, then to Δ'-pyrroline-5-carboxylate, to an unidentified symmetrical intermediate, and finally to the pyrrolidine ring of nicotine. This postulate would not be in conflict with the formation of the pyrrolidine ring from ornithine if there is δ-transamination of ornithine to glutamic semi-aldehyde. But if

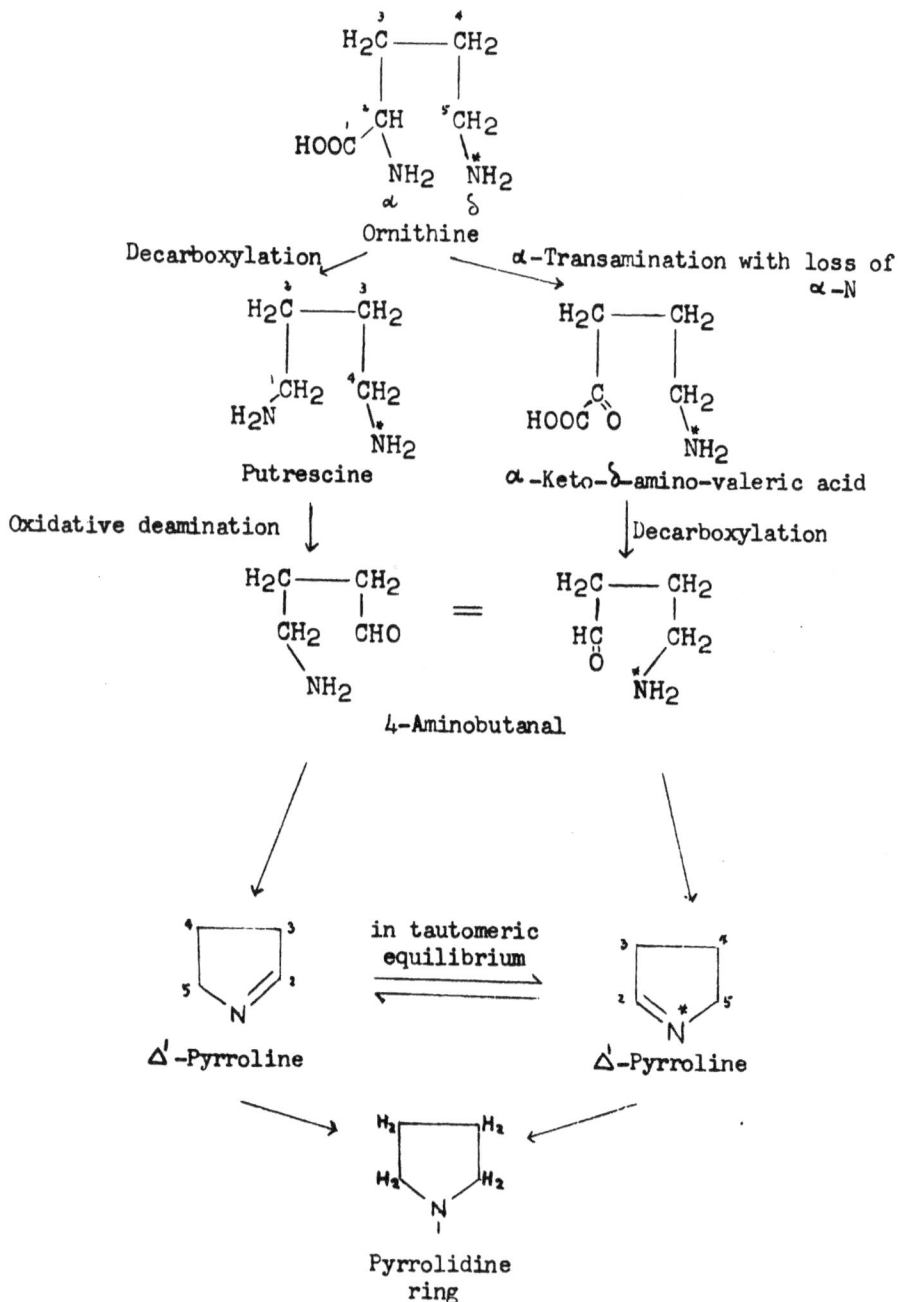

Figure 2. Postulated biogenesis of pyrrolidine ring from ornithine in Nicotiana

the formation of the pyrrolidine ring would occur through putrescine, 4-aminobutanal, and Δ'-pyrroline (see Leete's postulate in (e) above), the role of glutamic semi-aldehyde would be cast in some doubt.

(g) <u>Nicotiana glauca</u> normally produces about four parts of anabasine to one part of nicotine. When lysine-C^{14} or lysine-ε-N^{15} was fed to <u>N.</u> <u>glauca</u> excised root culture, the isotope became incorporated into the <u>piperidine ring</u> of the alkaloid anabasine, and very little incorporation of the isotope into nicotine occurred (1, 4, 10).

3.3 Alkaloids of Areca Nut

The dried ripe seed of <u>Areca Catechu</u> L. (Fam. Palmae), known as Areca Nut, or Betel Nut, contains several alkaloids, of which the principal ones are arecoline (about 0.1%) and arecaidine (in smaller quantity than arecoline). Their molecules contain a tetrahydropyridine nucleus.

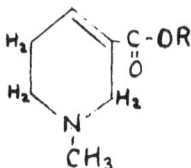

Arecoline   R = CH_3
Arecaidine   R = H

Arecoline and arecaidine are both oily liquids with boiling points at 209° C. and 223 - 224° C. respectively. Arecoline is volatile in steam, miscible with water and with most organic solvents. Its hydrobromide salt crystallizes from hot alcohol as slender prisms, and the platinichloride salt crystallizes from water as orange-red rhombs. Its aurichloride is an oil.

Arecaidine is soluble in water but not in most organic solvents. The platinichloride crystallizes in octahedra and the aurichloride in prisms, from hot dilute hydrochloric acid.

Arecoline is hydrolyzed by acid or alkali to arecaidine, and esterification of arecaidine with methanol yields arecoline.

Arecoline is used as an anthelmintic for animals for treatment of tape-worm or round-worm infestation, but is not so used for humans. However, it is of some pharmacological interest as it has parasympathomimetic action, stimulating those smooth muscles and exocrine gland cells innervated by post-ganglionic cholinergic nerve fibres.

Preparations in the National Formulary XII:

Areca

Arecoline Hydrobromide

3.4 References

1. Bothner-By, A., Dawson, R. F., & Christman, D. R. - Experientia, 12:151 (1956).

2. Bowden, K. - Nature, 172:768 (1953).

3. Dawson, R. F. - Adv. Enzymol., 8:203-251 (1948).

4. Dawson, R. F. - Am. Scientist, 48:321-340 (1960).

5. Dewey, L. J., Byerrum, R. U., & Ball, C. D. - Biochim. Biophys. Acta, 18:141-142 (1955).

6. Friedman, A. R., & Leete, E. - J. Am. Chem. Soc., 85:2141-2144 (1963).

7. Griffith, T., Hellman, K. P., & Byerrum, R. U. - Biochemistry, 1:336-340 (1962).

8. Grollman, A. - Pharmacology and Therapeutics, 6th ed., pp. 410-413, Lea and Febiger, Philadelphia, 1965.

9. Lamberts, B. L., Dewey, L. J., & Byerrum, R. U. - Biochim. Biophys. Acta, 33:22-26 (1959).

10. Leete, E. - Chem. and Ind., p. 537 (1955).

11. Leete, E. - Chem. and Ind., p. 1270 (1957).

12. Leete, E. - J. Am. Chem. Soc., 80:2162-2164 (1958).

13. Leete, E. - in P. Bernfeld (ed.): Biogenesis of Natural Compounds, p. 749, Pergamon Press, Oxford & London, 1963.

14. Leete, E. - Science, 147:1000-1006 (1965).

15. Leete, E., Gros, E. G., & Gilbertson, T. J. - Tetrahedron Letters, No. 11, pp. 587-592 (1964).

16. Solt, M. - Plant Physiol., 32:484-490 (1957).

17. Wu, P. H. L., Griffith, T., & Byerrum, R. U. - J. Biol. Chem., 237: 887-890 (1962).

CHAPTER 4

ALKALOIDS OF THE TROPANE GROUP

4.1 Introduction

The major tropane alkaloids (also called tropine alkaloids) of pharmacological or therapeutic importance may be grouped into (a) those occurrin in a number of species of the plant family Solanaceae, and (b) those occurring in certain Erythroxylon species (Fam. Erythroxylaceae).

4.2 The Solanaceous Alkaloids

(a) Plant Sources

A number of species of the genera Atropa, Datura, Hyoscyamus, Duboisia, and Scopolia, all belonging to the Solanaceae family, produce in varying quantities the tropane alkaloids hyoscyamine (of which the racemic form is atropine), scopolamine (also called hyoscine), meteloidine (3-tiglylteloidine), 3,6-ditiglylteloidine, belladonnine, and several other alkaloids chemically related to these. Of these, hyoscyamine and scopolamine are the major ones both in the sense that they occur in these species in much greater quantities than the other alkaloids and in that they are pharmacologically important drugs. All of these alkaloids are often collectively referred to as the "solanaceous alkaloids" although there are other alkaloids which are found in the Solanaceae family and which are traditionally not referred to as "solanaceous alkaloids". These "solanaceous alkaloids" are sometimes also referred to as the "Belladonna alkaloids", although they occur in a great many plant species other than Atropa Belladonna, as will be seen below.

Listed below are some of the better known species which bear these solanaceous alkaloids, and the average alkaloidal content as per cent of dried plant material (2, 5, 9, 21). There are considerable variations in alkaloidal content among different plants of the same species, and also variations of alkaloidal content, in many cases, between younger and older plants. However, in the fully mature plant, certain particular species always accumulate a greater proportion of hyoscyamine than scopolamine, while certain other species accumulate more scopolamine than hyoscyamine. Unless otherwise indicated, the data listed below refer to the mature plant in each case. As will be noted, in many cases, data for the alkaloidal content of the different parts of the plant are not available in the literature. Although as crude drug (dried plant material), only Atropa Belladonna, Datura stramonium, and Hyoscyamus niger are recognized and specified for certain preparations in the U.S.P., N.F. and B.P.C., many of the other species listed below also contain these alkaloids to nearly the same extent, and therefore can also serve as raw materials for the extraction and isolation of these alkaloids where and when these other species are available.

From the data given in Table 2, it will be noted that certain Datura species such as D. stramonium, D. tatula, the Atropa species and the Hyoscyamus species accumulate more hyoscyamine than scopolamine. On the other hand, certain other Datura species such as D. metel, D. ferox, D. fastuosa, D. arborea, D. innoxia, D. inermis, etc., accumulate more scopolamine than hyoscyamine.

Table 2. Principal Solanaceae Plants Bearing Tropane Alkaloids*

Plant Species	Alkaloid	Plant Part and Alkaloid Content (%)
Atropa acuminata Royle (Indian Belladonna)	Total (mostly hyoscyamine)	(W) 0.3-0.5
Atropa Belladonna L. (Deadly Nightshade)	Total (mostly hyoscyamine; some scopolamine)	(L) 0.3-0.4; (R) 0.4-0.6; (Sd) 0.3-0.8; (F) 0.2
Datura arborea L.	Scopolamine Hyoscyamine	(W) 0.4 (Young St. & R) trace
Datura fastuosa L.	Scopolamine Hyoscyamine	(L) 0.1 (L) 0.02 or less
Datura ferox L.	Scopolamine Meteloidine 3,6-ditiglylteloidine	(W) 0.3 (W) 0.1 (R) 0.05
Datura inermis Jacq.	Hyoscyamine Scopolamine	(W) 0.04-0.07 (W) 0.1-0.2
Datura innoxia Mill.	Scopolamine Hyoscyamine	(W) 0.3 (W) 0.06
Datura leichhardtii Muell. ex Benth	Total (Hm. and Sc. in about equal amounts in younger plants, and less Sc. than Hm. in older plants)	(aerial parts) 0.15
Datura metel L.	Scopolamine Hyoscyamine	(W) 0.1 (W) 0.04
Datura meteloides DC	Scopolamine Hyoscyamine Meteloidine	(W) 0.1 (W) 0.03 (W) 0.05
Datura quercifolia HBK	Hyoscyamine	(L) 0.4; (Sd) 0.3

Table 2 (Continued)

Plant Species	Alkaloid	Plant Part and Alkaloid Content (%)
Datura sanguinea Ruiz et Pavon	Hyoscyamine Scopolamine	(aerial parts) 0.02; (R) 0.4 (aerial parts) 0.35; (R) 0.2
Datura stramonium L. (Jimson Weed) (Jamestown Weed)	Total (mostly hyoscy- amine)	(L) 0.2-0.5; (R) 0.2-0.25; (Sd) 0.2-0.5
Datura tatula L.	Total Hyoscyamine Scopolamine 3,6-ditiglylteloidine	(L) 0.2-0.4; (W) 0.15-0.3; (W) 0.07; (R) 0.01
Duboisia myoporoides R. Br.	Total (proportion of hyoscyamine and scopolamine varies)	(L) 0.9-4.0
Hyoscyamus albus L.	Total (mostly hyoscyamine)	(L) 0.2-0.56
Hyoscyamus muticus L. (Egyptian Henbane)	Total (mostly hyoscyamine)	(L) 0.5-1.4
Hyoscyamus niger L. (Henbane)	Total (mostly hyoscyamine)	(L) 0.04-0.08; (Sd) 0.1-0.15
Scopolia carniolica Jacq.	Total	Rhizome 0.4-0.6
Scopolia japonica Maxim (Japanese Belladonna)	Hyoscyamine	(L) 0.18

(W) = whole plant   (L) = leaves
(R) = roots   (Sd) = seeds
(St) = stems   (F) = fruits
(Hm) = hyoscyamine   (Sc) = scopolamine

*Adapted in part by permission from A. G. Avery, S. Satina and J. Rietsema – BLAKESLEE: THE GENUS DATURA, Copyright 1959, The Ronald Press Co., New York, and from references 2, 5, 9, and 21.

(b) <u>Properties of the Principal Alkaloids</u>

In these Solanaceae plants listed above, the major alkaloids of medicinal importance are hyoscyamine, atropine, and scopolamine. Hyoscyamine and scopolamine (also called hyoscine) are both laevorotatory. Atropine (the racemic form of hyoscyamine) and atroscine (the racemic form of scopolamine) normally do not occur in the plants in more than trace quantities, if at all. But racemization of (-)-hyoscyamine and of (-)-scopolamine to their respective racemic forms often occurs during extraction. Racemization of hyoscyamine and scopolamine to atropine and atroscine respectively may be effected by addition of small quantity of caustic alkali, or sodium carbonate or ammonia to their cold alcoholic solution or by heating their chloroform solution.

<u>Hyoscyamine and Atropine</u>

Hyoscyamine, the free base, is readily soluble in chloroform (1 gm. in 1 ml.), in alcohol or in benzene, a little less readily soluble in ether (1 gm. in 69 ml. at 25°), and only sparingly soluble in water (1 gm. in 281 ml. at 25°). Its ordinary salts which are readily soluble in water but nearly insoluble in chloroform or ether, are crystalline. Hyoscyamine sulphate $(B)_2.H_2SO_4.2H_2O$ crystallizes in needles from alcohol, and its aurichloride $(B).HAuCl_4$ crystallizes in yellow hexagonal plates from dilute hydrochloric acid, in which it is less soluble than atropine aurichloride. Hyoscyamine hydrobromide, which is non-deliquescent, forms prisms.

Atropine is readily soluble in alcohol (1 gm. in 2 ml.), in chloroform (1 gm. in 1 ml.), and in ether (1 gm. in 25 ml.). It is only sparingly soluble in water (1 gm. in 460 ml.) and practically insoluble in petroleum ether. Atropine crystallizes from alcohol on addition of water, or from chloroform on addition of petroleum ether. It may be sublimed unchanged by rapid heating.

Atropine sulphate $(B)_2.H_2SO_4.H_2O$ is the most commonly used form among its salts for medicinal purposes. It is readily soluble in water (1 gm. in 0.5 ml.) and in alcohol (1 gm. in 5 ml.), but only sparingly soluble in ether (1 gm. in 2140 ml.). Atropine sulphate crystallizes from its alcoholic solution by addition of acetone. There are also other crystalline atropine salts such as the hydrobromide, oxalate, aurichloride and platinichloride. Some of these crystalline salt forms serve as an aid to identification of the alkaloid. Atropine methyl nitrate and atropine methyl bromide are also used as medicinal agents.

A number of minor alkaloids occurring in trace quantities in certain Atropa, Datura, and Hyoscyamus species include apoatropine and belladonnine. Apoatropine $C_{17}H_{21}O_2N$ is the anhydride of atropine $C_{17}H_{23}O_3N$. Belladonnine is thought to be a dimeride $(C_{17}H_{21}O_2N)_2$ related to apoatropine or its isomer.

Scopolamine (Hyoscine)

Scopolamine (the free base) is a syrupy, laevorotatory substance, soluble in chloroform and alcohol, but only sparingly soluble in benzene or petroleum ether. Its salts are readily soluble in water and in alcohol, sparingly soluble in chloroform and insoluble in ether. Scopolamine hydrobromide, the most commonly used form among the scopolamine salts for medicinal purposes, forms rhombic tablets. Other crystalline salts of scopolamine include the aurichloride and the picrate, both of which may be used to distinguish this alkaloid from the other Solanaceous alkaloids.

Hydrolysis

Heating hyoscyamine or atropine with acids or alkalis hydrolyzes the alkaloid to tropine and tropic acid. Scopolamine is hydrolyzed by heating with dilute acid or alkali to tropic acid and scopoline. Scopolamine is also slowly hydrolyzed by pancreatic lipase (in ammonia-ammonium chloride buffer) to tropic acid and scopine. Scopine may be distinguished from scopoline by the different crystalline characters of their aurichlorides. Scopine is converted by acid or base to scopoline.

Meteloidine

Meteloidine occurs in Datura meteloides and certain other species of Datura. It is soluble in alcohol and in chloroform, and sparingly soluble in water, ether or benzene. Its salts, the hydrochloride and hydrobromide, are crystalline. On hydrolysis, meteloidine yields teloidine, and tiglic acid. Meteloidine is pharmacologically inactive.

Colour Reactions

Hyoscyamine, atropine, and scopolamine all give the Vitali-Morin colour reaction (7). This reaction may be tested in this way: A minute quantity (as little as 1 μg.) of the solid alkaloid and a drop of fuming nitric acid are mixed in an evaporating dish, and evaporated to dryness at 100° C. To the residue is then added 0.5 ml. of a 3% solution of potassium hydroxide in methanol (freshly prepared). A bright purple colour is produced which changes to red and subsequently fades to colourless. Under standard-ized conditions of volume and concentration limits and time interval, the intensity of the colour produced by this reaction may also serve as the

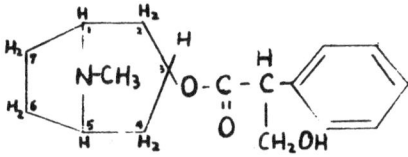

Hyoscyamine
(Atropine)

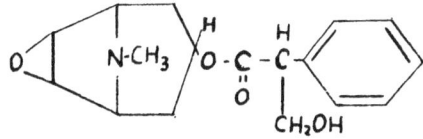

Scopolamine

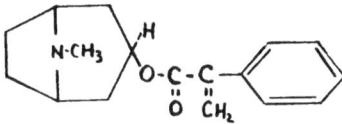

Apoatropine

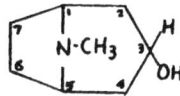

Tropine

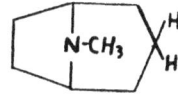

Tropane

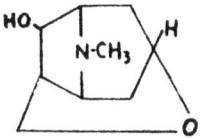

Scopoline

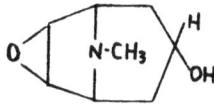

Scopine

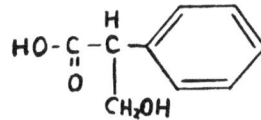

Tropic Acid

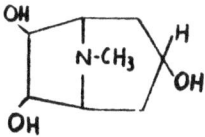

Teloidine

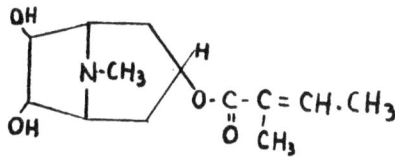

Meteloidine

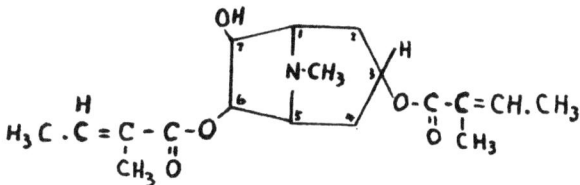

3,6-ditiglylteloidine

basis of quantitative estimation of the total alkaloids extracted from plant samples (7).

Hyoscyamine and atropine also give the <u>Gerrard reaction</u>, in which a few mg. of the alkaloid react with a 2% solution of mercuric chloride in 50% alcohol to produce a red colour without warming (for atropine) or upon warming (for hyoscyamine). Scopolamine gives a white precipitate with mercuric chloride.

(c) <u>Assay for Total Alkaloids</u>

The procedure of the U.S.P. assay for Belladonna Leaf (24) can be adapted for quantitative estimation of total alkaloids in plant materials of other species containing hyoscyamine and scopolamine as principal alkaloids. In this procedure, the alkaloids are extracted by ether, purified by re-extracting into 0.5 N. sulphuric acid (as the sulphates) and then into chloroform (as the free bases). After evaporating the chloroform extract to dryness, the alkaloidal residue is taken up with a definite quantity of standardized sulphuric acid solution in slight excess of the quantity of acid required to form sulphate salts with all the alkaloids present. The quantity of the unreacted acid is determined by titration with standardized alkali, and the quantity of the alkaloids is calculated from the molar quantity of the acid which has reacted with the alkaloids to form the salts $B_2.H_2SO_4$. When the molecular weight of hyoscyamine is used in such calculations, the results are expressed as quantity (or per cent) of "total alkaloids calculated as hyoscyamine", as is usually done.

(d) <u>Isolation</u>

For the isolation of the individual alkaloids, the alkaloidal residue left after the chloroform has been removed by distillation (in the assay procedure) may be neutralized with oxalic acid, and the oxalates of atropine and hyoscyamine may be separated by fractional crystallization from acetone and ether (in which hyoscyamine oxalate is more soluble than atropine oxalate).

(e) Action and Uses

Hyoscyamine, atropine, and scopolamine act on tissue cells innervated by post-ganglionic cholinergic fibres of the parasympathetic nervous system so that the response of these effector cells (and the organs containing them) to the parasympathetic nerve impulses is inhibited or abolished. They are therefore parasympatholytic or cholinergic-blocking in action. Consequently they relax bronchial and intestinal smooth muscles (antispasmodic action), inhibit contraction of the iris muscles of the eye to produce mydriasis, and decrease salivary and sweat gland secretions. In addition, scopolamine has some sedative effect on the central nervous system. Atropine has some initial stimulating action on medullary centres of the central nervous system and this is followed by depression or paralysis.

These alkaloids and the plant drugs containing them are often used (sometimes in combination with barbiturate or other sedative agents) in pre-operative medication (to inhibit secretions), for mydriasis, and for diarrhea. The usual therapeutic dosage range for atropine is 0.25-1 mg. The dosage ranges for the plant drugs (leaf or root), the tinctures, and other pharmaceutical preparations correspond to this dosage for atropine in accordance with their alkaloidal contents. It should be pointed out that (+)-hyoscyamine (i.e., d-hyoscyamine) is pharmacologically nearly inactive.

Atropine being the racemic form of hyoscyamine has pharmacological action equivalent to only about one-half that of (-)-hyoscyamine (i.e., l-hyoscyamine) of the same weight. When applied directly to the conjunctival sac of the eye (cat, dog, and rabbit), 0.5 µg. of atropine sulphate in solution is sufficient to cause dilation of the pupil in one hour.

(f) <u>Medicinal Preparations</u>

Principal preparations in U.S.P. XVII, N.F. XII, B.P. 1963, and B.P.C. 1963 containing these Solanaceous plant materials or their alkaloids are the following:

Belladonna Leaf U.S.P., B.P.C.

Belladonna Root B.P.C.

Belladonna Extract N.F., B.P.

Belladonna Leaf Fluid Extract N.F.

Belladonna Liquid Extract B.P.

Belladonna Tincture U.S.P., B.P.

Hyoscyamus B.P.C.

Hyoscyamus Extract B.P.

Hyoscyamus Tincture N.F., B.P.

Stramonium B.P.C.

Stramonium Dry Extract B.P.C.

Stramonium Liquid Extract B.P.

Stramonium Tincture B.P.

Atropine N.F., B.P.C.

Atropine Sulphate U.S.P., B.P.C.

Atropine Methonitrate B.P.C.

Hyoscyamine Sulphate N.F.

Scopolamine Hydrobromide U.S.P., (Hyoscine Hydrobromide B.P.C.)

Examples of Pharmaceutical Specialties:

Bardase[R]; Bellafoline[R]; Butibel[R]; Delkadon[R]; Donnatal[R];

Levsinex[R]; Nembu-Donna[R]; Prydon[R]; Rabellon[R]; Wyanoid[R].

(g) Biosynthesis of Tropane Alkaloids

Numerous experimental studies have been made in the past decade on the biosynthetic formation of the tropane portion, the N-methyl group, and the tropic acid portion of hyoscyamine and scopolamine.

1. Biosynthesis of the tropane portion - Based on chemical considerations, Robinson's proposal (19) of three decades ago suggested the biosynthetic route from ornithine and arginine through succinic dialdehyde (derivable from carbohydrates) and tropanone to tropine. This proposal and modifications of it have served as the basis or working hypothesis for many experimental studies by various investigators. Experimental results have now confirmed some of these precursors and modified others in the original Robinson scheme.

The most significant experimental investigations and the conclusions drawn from their results may be summarized as follows:

(a) Cromwell (1) and James (6) fed detached leaves or injected intact plants of Atropa and Datura with arginine, ornithine, or putrescine, and found increased hyoscyamine content. Reinouts (18) fed putrescine to isolated roots of Atropa belladonna in sterile media, and found an increase of growth and an increase in total alkaloid content without affecting the ratio of the different alkaloids.

(b) Leete et al. (11, 14) fed ornithine-2-C^{14} to Datura stramonium, and found that the C^{14} isotope was incorporated into one of the two bridge-head carbon atoms (C_1 or C_5) of the tropine part of hyoscyamine. The isotope from ornithine-2-C^{14} was also so incorporated into scopolamine by young plants (two-month old) (11, 22) but not by the older (five-month old) plants (14). There is some evidence that scopolamine is formed in the earlier stage and not in the fully mature plant.

(c) Incorporation of putrescine-1,4-C^{14} into both hyoscyamine and scopolamine was shown by Kaczkowski and Marion (8) using root cultures of Datura metel. Leete and Louden (13) using intact Datura stramonium plants also found incorporation of putrescine-1,4-C^{14} into hyoscyamine.

The conclusion is that ornithine is incorporated into the tropine portion of these alkaloids **not via** putrescine, and putrescine's incorporation into the alkaloid while possible does not represent the normal route in the plant. For, if ornithine-2-C^{14} were incorporated via putrescine, both C_1 and C_5 of the tropine would be found labelled with the isotope, and not just one of these two carbon atoms as the experimental results show. Ornithine-2-C^{14} would give rise to putrescine-1-C^{14} (by decarboxylation) which would then result in both C_1 and C_5 of the tropine being labelled because putrescine is a symmetrical molecule and C_1 and C_4 of putrescine would be indistinguishable in metabolic reactions.

(d) Conversion of hyoscyamine to scopolamine has been shown in different species of Datura (20). This is considered to occur via an intermediate 6,7-dehydro-hyoscyamine (i.e., hyoscyamine with a double bond between C_6 and C_7).

(e) A number of compounds have been suggested as possible or probable intermediates through which ornithine may proceed to the formation of the tropine portion of these alkaloids (12). Experimental proof of these remains to be fully established.

2. Biosynthetic source of the N-Methyl Group - Feeding methionine-$C^{14}H_3$ by Marion and Thomas (16) to mature (five-month old) Datura stramonium plants resulted in the incorporation of the isotope into the N-methyl group in hyoscyamine.

The S-methyl group of methionine has also been shown to contribute to the N-methyl group in other alkaloids (e.g., in the ergot alkaloids).

3. Biosynthesis of the tropic acid portion - Leete (10) and Underhill and Youngken (23) fed phenylalanine-3-C^{14} to Datura stramonium and found the isotope to be incorporated into the tropic acid portion (including the carboxyl group) of the alkaloid. When phenylacetic acid-1-C^{14} was fed to the plant (23), the isotope was incorporated into the styrene part of the tropic acid but not the carboxyl group. Incorporation of phenylalanine-3-C^{14} into the tropic acid portion of hyoscyamine was also shown in Datura metel by Gross and Schutte (4).

Phenylacetic acid in its participation in the biosynthesis of the tropic acid part of the alkaloid is assumed to be in the form of phenylacetyl-Co.A, which may conceivably proceed through phenylmalonyl-Co.A to tropic acid (15). The metabolic origin of the hydroxymethyl ($-CH_2OH$) group of the tropic acid remains uncertain.

4.3 Alkaloids of Erythroxylon Coca

The leaves of Erythroxylon coca Lamarck (Fam. Erythroxylaceae) and

Erythroxylon truxillense Rusby, which are tropical trees, contain several
alkaloids of which cocaine is the major and best known one. The alkaloidal
contents of these leaves usually range from 0.2 to 1 per cent, but higher
total alkaloidal contents have been reported. The proportion of cocaine in
the total alkaloid content varies from fifty to ninety per cent.

Cocaine is benzoylmethylecgonine. In addition to cocaine, the
several other alkaloids occurring in these species include other ecgonine
derivatives as well as a number of alkaloids related to hygrine (which contains pyrrolidine nucleus). Ecgonine is tropine-2-carboxylic acid.

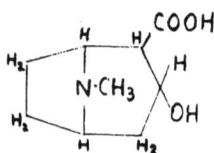

Ecgonine

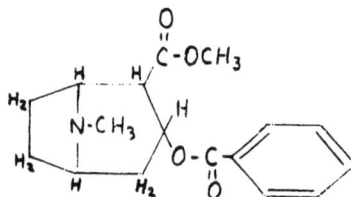

Cocaine

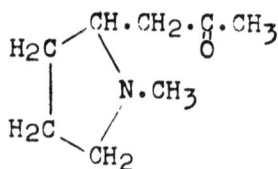

Hygrine

Cocaine, a laevorotatory tertiary amine, is slightly soluble in
water (1 gm. in 625 ml. at 25° C.), and readily soluble in alcohol (1 gm. in
5 ml.), ether (1 gm. in 3.9 ml.), benzene, chloroform, or petroleum ether.
It crystallizes from alcohol in prisms.

Cocaine hydrochloride is readily soluble in water (1 gm. in 0.5 ml.)

or alcohol (1 gm. in 2.6 ml.), but insoluble in ether or petroleum ether. It crystallizes from alcohol. In micro-precipitation procedures (3, 17), cocaine or its salt forms precipitates with a number of alkaloid-precipitating reagents such as 5% gold chloride, Mayer's, and Wagner's reagents. Some of these precipitates form crystalline aggregates which are sometimes utilized as aid to confirmation of identity of the alkaloid by microscopic examination.

Addition of a drop of saturated solution of potassium permanganate to a solution of cocaine in 50%-saturated solution of alum gives a crystalline precipitate with characteristic crystal forms.

Cocaine is hydrolyzed by acids to ecgonine, benzoic acid, and methyl alcohol. Ecgonine forms salts with both bases and acids.

Actions

Cocaine possesses the ability to block nerve conduction upon local application (local anaesthetic action). Its systemic effect is stimulation of the central nervous system but this action is not used for clinical purposes as the alkaloid is too toxic in its several other side actions. There is also the danger of habit formation. For local anaesthetic purposes, a number of synthetic drugs are now generally preferred. Solutions used for surface (local) anaesthesia vary from 1% to 10%, depending on the mucosa to be anaesthetized. Severe toxic symptoms from systemic effect have been reported from as little as 20 mg. of cocaine.

Medicinal Preparations

Cocaine N.F. XII, B.P.C. 1963

Cocaine Hydrochloride U.S.P. XVII, B.P.C. 1963

4.4 References

1. Cromwell, B. T. - Biochem. J., 37:722-726 (1943).

2. Evans, W. C., & Stevenson, N. A. - J. Pharm. Pharmacol., 14:107T (1962).

3. Farmilo, C. G., & Genest, K. - in C. P. Stewart and A. Stolman (eds.):
    Toxicology: Mechanism and Methods of Analysis, Vol. II, pp. 229-242,
    Academic Press, New York, 1961.

4. Gross, D., & Schutte, H. R. - Arch. d. Pharm., 296:1-6 (1963).

5. Henry, T. A. - The Plant Alkaloids, 4th ed., p. 65, Blakiston, Phila-
    delphia, 1949.

6. James, W. O. - New Phytologist, 48:172-185 (1949).

7. James, W. O., & Roberts, M. - Quart. J. Pharm., 18:29-35 (1945);
    20:1-16 (1947).

8. Kaczkowski, J., & Marion, L. - Can. J. Chem., 41:2651-2653 (1963).

9. Leete, E. - in A. G. Avery, S. Satina, & J. Rietsema (eds.): Blakeslee:
    The Genus Datura, pp. 48-56, Ronald Press, New York, 1959.

10. Leete, E. - J. Am. Chem. Soc., 82:612-614 (1960).

11. Leete, E. - J. Am. Chem. Soc., 84:55-57 (1962).

12. Leete, E. - in P. Bernfeld (ed.): Biogenesis of Natural Compounds,
    p. 745, Pergamon Press, Oxford & London, 1963.

13. Leete, E., & Louden, M. C. L. - Chem. and Ind., p. 1725 (1963).

14. Leete, E., Marion, L., & Spenser, I. D. - Can. J. Chem., 32:1116-1123
    (1954).

15. Louden, M. C. L., & Leete, E. - J. Am. Chem. Soc., 84:1510-1511 (1962).

16. Marion, L., & Thomas, A. F., - Can. J. Chem., 33:1853-1854 (1955).

17. Martello, R., & Farnsworth, N. R. - Lloydia, 25:176-185 (1962).

18. Reinouts van Haga, P. - Biochim. Biophys. Acta, 19:562 (1956).

19. Robinson, R. - J. Chem. Soc., 111:876-899 (1917); pp. 1079-1090 (1936).

20. Romeike, A., & Fodor, G. - Tetrahedron Letters, No. 22, pp. 1-4 (1960).

21. Scott, W. E., Ma, R. M., Schaffer, P. S., & Fontaine, T. D. - J. Am. Pharm. Assoc., Sci. Ed., 46:302-304 (1957).

22. Turner, F. A., & Gearien, J. E. - J. Pharm. Sci., 53:1309-1312 (1964).

23. Underhill, E. W., & Youngken, H. W. Jr. - J. Pharm. Sci., 51:121-125 (1962).

24. United States Pharmacopeia, 17th Rev., p. 62, Mack Publishing Co., Easton, Pa., 1965.

CHAPTER 5

ALKALOIDS OF THE ISOQUINOLINE GROUP

5.1 Introduction

Alkaloids which chemically belong to various sub-groups of the iso-quinoline group and which are of pharmacological and therapeutic importance, include the alkaloids in Opium, in Ipecac, and in Curare.

5.2 Alkaloids of Opium

When incisions are made on the unripe fruit (botanically a capsule) of the opium poppy, Papaver somniferum L. (Fam. Papaveraceae), a milky exudate soon collects around these incisions. As it is air-dried, this exudate darkens to a greyish-brown colour, and hardens a little to become clay-like in consistency. It becomes hard and brittle on further storage. This dried mass is called Opium. It has a characteristic odour and a bitter taste.

More than twenty alkaloids have been isolated from opium. Among them, the six or seven major alkaloids of pharmacological and therapeutic importance may be grouped into two groups: (a) morphine and related alkaloids which contain a structural nucleus related to phenanthrene, and (b) those alkaloids which contain a benzylisoquinoline nucleus or its derivative.

(a) Morphine Group

Morphine, codeine, and thebaine are the major ones in this group. These occur in opium to the extent of 8 - 15% for morphine, 0.8 - 2.5% for codeine, and 0.1 - 1% for thebaine. These percentage contents refer to unstandardized raw opium, and not Opium of the U. S. Pharmacopeia or British Pharmacopeia which specify a content of not less than 9.5% morphine. As the capsule (fruit) ripens, the morphine content decreases.

Ethylmorphine, diacetylmorphine (heroin), dihydromorphinone (hydromorphone, dilaudid), and dihydrocodeinone (hydrocodone) do not occur in opium
as such, but are prepared by chemical means from the naturally occurring alkaloids. These derivatives are pharmacologically active and important.

Morphine (the free base), unlike most other alkaloids in their free-
base forms, is only sparingly soluble in chloroform, and nearly insoluble in
ether or benzene. Its solubility in various solvents differs rather markedly
from the solubility of codeine, and this difference can be utilized in their
separation. The solubility data (at 25° C. unless otherwise indicated) for
morphine, codeine, and some of their salts are listed in Table 3.

Table 3. Solubility Data of Morphine,
Codeine, and Their Salts

Alkaloid or its salt	ml. of solvent to dissolve 1 gm. of alkaloid or alkaloid salt at 25° C.					
	Water	Alcohol	Ether	Chloroform	Ammonia	NaOH or KOH soln.
Morphine	3400	300	5000	1525	120	100
Codeine	120	2	50	0.5	68 (15.5°)	nearly insol.
Morphine Sulphate $(B)_2 \cdot H_2SO_4 \cdot 5H_2O$	15.5	565	insol.			
Codeine Sulphate	30	1280	insol.			
Codeine Phosphate	2.5	325	insol.			

Morphine contains two hydroxyl groups, of which the one at carbon

No. 3 is phenolic and the one at carbon No. 6 is a secondary alcohol group.
Morphine is a monoacidic base with pK_a of 9.85. The average pH of morphine
salts in solution is 4.68 and therefore methyl red may be used as indicator
for titrating the base. Morphine is laevorotatory; so are its sulphate and
hydrochloride.

When morphine or its hydrochloride is heated at 140° C. under pressure (in sealed tube) with hydrochloric acid, apomorphine is formed.

Codeine, which is the methyl ether of morphine, is a monoacidic base
with pK_a of 7.95 (at 15°). The pH of codeine salts is similar to that of
morphine salts (e.g., codeine hydrochloride, pH 4.93). The rather marked
difference between the solubility of codeine and that of morphine in various
solvents has already been pointed out. In addition to the data given in
Table 3, it may also be noted that codeine is soluble in cold benzene (1 gm.
in 13 ml.) whereas morphine is not. Codeine is sparingly soluble or nearly
insoluble in alkali hydroxide, while morphine is fairly soluble in alkali
hydroxide (1 gm. in 100 ml.). On treatment with hydrochloric acid, codeine
also yields apomorphine accompanied by other products.

Thebaine is readily soluble in alcohol, chloroform, or benzene,
sparingly soluble in lime-water or ammonia, and nearly insoluble in water.
It may be separated from the other major alkaloids of opium as the salicylate, which is only sparingly soluble in water.

Colour and Precipitating Tests

In micro-precipitating tests (4, 6, 12), morphine and codeine give
crystalline precipitates of more or less characteristic forms when tested
with certain reagents such as Marme's and Wagner's reagents. With certain

- 56 -

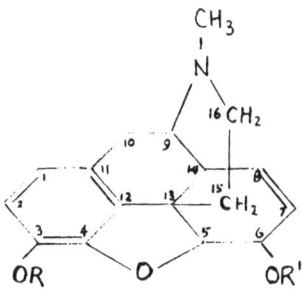

	R =	R' =
Morphine	H	H
Codeine	CH_3	H
Ethylmorphine	C_2H_5	H
Heroin	$CH_3 \cdot \overset{O}{\underset{\parallel}{C}} -$	$CH_3 \cdot \overset{O}{\underset{\parallel}{C}} -$

Dihydro-morphinone   H   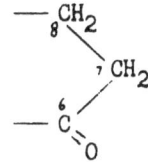

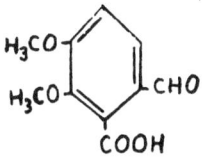

Opianic acid

Dihydro-codeinone CH_3 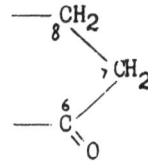

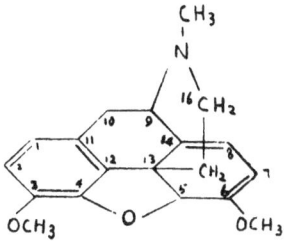

Thebaine

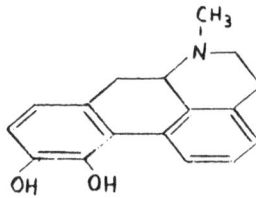

Apomorphine

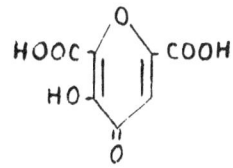

Meconic acid

other reagents, these alkaloids also produce colour reactions. A few of these reagents are listed in Table 4. These tests are not sufficiently specific and reliable for use by themselves for identification of the alkaloids, but are sometimes used as supporting evidence when other tests are carried out.

Table 4. Some Colour Reactions of
the Opium Alkaloids

Alkaloid	Selenium dioxide 5% solution	Marquis reagent*	Neutral ferric chloride solution	Nitric acid
Morphine	blue-green ↓ ** grey-green	violet	blue (fading on warming or adding acid or alcohol)	faintly pink
Codeine	blue-green ↓ yellow-green ↓ brown	violet	no colour	yellow
Thebaine	green ↓ brown ↓ orange	red ↓ orange		colour-less ↓ yellow
Papaverine	grey-green ↓ fading	blue ↓ violet ↓ green-brown		
Noscapine	green ↓ orange	blue-violet ↓ rapidly fading		

*Concentrated sulphuric acid containing one drop of 40% formaldehyde per ml.
**Arrows denote sequence of colour changes.

Meconic Acid

Beside the alkaloids, Opium also contains meconic acid (3 - 5%) some of which may be in combination with the alkaloids as meconate. A positive test for the presence of meconic acid is sometimes taken to indicate possible presence of opium in a powdered sample. The test for meconic acid may be carried out by warming 20 - 30 mg. of powdered opium in 2 or 3 ml. of water for a few minutes, and then filtering. On adding a few drops of 5% ferric chloride solution to the filtrate, a purplish red colour is produced, and the colour is not destroyed by addition of dilute hydrochloric acid or 5% mercuric chloride solution.

(b) Benzylisoquinoline Alkaloids in Opium

In addition to the morphine group of alkaloids, there also occur in Opium several alkaloids which are benzylisoquinoline derivatives. The major ones in this latter group are papaverine (0.5 - 1.5% in Opium), noscapine (4 - 8%), and narceine (0.1 - 0.7%). Noscapine was formerly known as narcotine. These benzylisoquinoline alkaloids in Opium are chemically as well as pharmacologically distinct from the morphine group of alkaloids. In their biosynthesis, however, both groups may be derived from tyrosine as their precursor. Reticuline, a new alkaloid recently isolated from the opium poppy (1), also has a structure related to this group.

Papaverine is a weak base, behaving as a tertiary amine. It is insoluble in water, slightly soluble in cold alcohol or ether, and soluble in chloroform. It dissolves in pure sulphuric acid to form a colourless solution, which becomes red at 110° C., and the colour is discharged by addition of water.

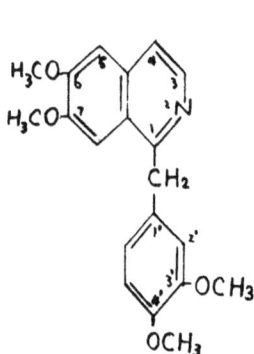

Papaverine

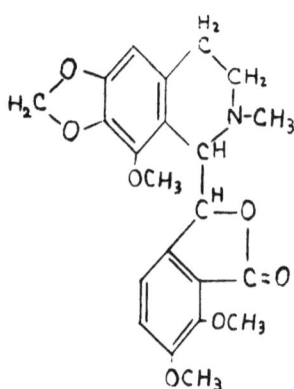

Noscapine

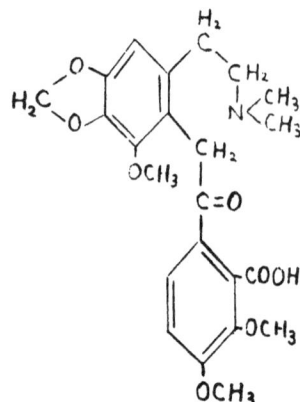

Narceine

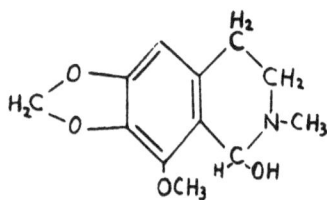

Cotarnine

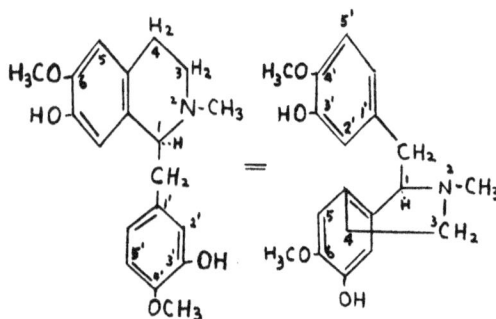

Reticuline

Noscapine is a weak monoacidic tertiary base. It is nearly insoluble in water, sparingly soluble in cold alcohol or ether, and readily soluble in benzene, acetone, or ethyl acetate. It is insoluble in cold alkali or ammonia, but soluble in hot alkali. It forms unstable salts with acids, and these salts are dissociated by water. When boiled with dilute acids, noscapine is hydrolyzed to opianic acid and hydrocotarnine.

Narceine occurs in Opium as such, but it can also be prepared from noscapine. Narceine behaves as a weak monoacidic tertiary base. It yields well-crystallized salts such as hydrochloride, picrate, and aurichloride. Narceine dissolves in ammonia, and in solutions of alkali hydroxides, forming crystallizable metallic derivatives. It reacts with phenylhydrazine or hydroxylamine to give the phenylhydrazone or oxime. It also esterifies with alcohols in the presence of hydrogen chloride.

(c) Separation of the Major Alkaloids from Opium

Largely on the basis of the properties described above, the six major alkaloids may be separated from Opium by a procedure (5b) represented by the schematic outline shown in Figure 3.

(d) Biosynthesis of Opium Alkaloids

In the past decade, experimental investigations with the use of radioactive isotope labelled compounds have given support to certain hypothetical schemes for the biosynthetic route of these alkaloids. Some of the most significant results may be summarized as follows:

(i) Battersby and Harper (2) administered tyrosine-2-C^{14} to Papaver somniferum and isolated C^{14}-labelled papaverine from it. By degradation of the papaverine so obtained, they found the C^{14}-label to be incorporated into positions C-1 and C-3 of the papaverine molecule in essentially equal amounts. This is taken to show that the benzylisoquinoline structure of the papaverine molecule originates from two molecules of tyrosine through certain appropriate intermediate compounds which are of fairly common occurrence in various known biological systems (e.g., 3,4-dihydroxyphenylalanine), and probably via nor-laudanosoline.

(ii) Administration of tyrosine-2-C^{14} to the opium poppy also resulted in the incorporation of the isotope label into morphine and codeine

Powdered Opium
↓
Shake with warm calcium chloride solution
↓
Filter

Insoluble matter   Filtrate
(discard)   (hydrochlorides of alkaloids)
↓
Reduce volume (evaporate under
reduced pressure) to syrupy liquid

Add 10% NaOH solution

Precipitates   Alkaline solution
(noscapine, papaverine, thebaine)   (morphine, codeine, narceine)
↓   ↓
Dissolve in dilute alcohol   Extract with chloroform
↓
Add acetic acid to make slightly acidic
↓   Chloroform   Aqueous
Add 3 volumes of boiling water   extract   alkaline
                                      (containing   solution
Precipitate   Solution   codeine)   (morphine,
(papaverine,   (thebaine)   ↓   narceine)
noscapine)   ↓   Further   ↓
                     Further   purification
                     purification   Make acidic
                                                         ↓
                                                         Aqueous acidic
Dissolve in boiling 0.33%   soln. (salts of
(aqueous) oxalic acid soln.   morphine and
↓   narceine)
Allow to stand

          Solution   Make slightly
          ↓   alkaline with
          Bring to boiling } Repeat   ammonia
          ↓
          Allow to stand   }   Precipitate   Solution
                                     (morphine)   (narceine)
Crystals   Solution   ↓   ↓
(papaverine   (noscapine   Further   Further
acid oxalate)   acid oxalate)   purification   purification
                      ↓
          Make alkaline with ammonia

Precipitate   Solution
(noscapine)   (discard)
↓
Dissolve in boiling alcohol
↓
Crystallization

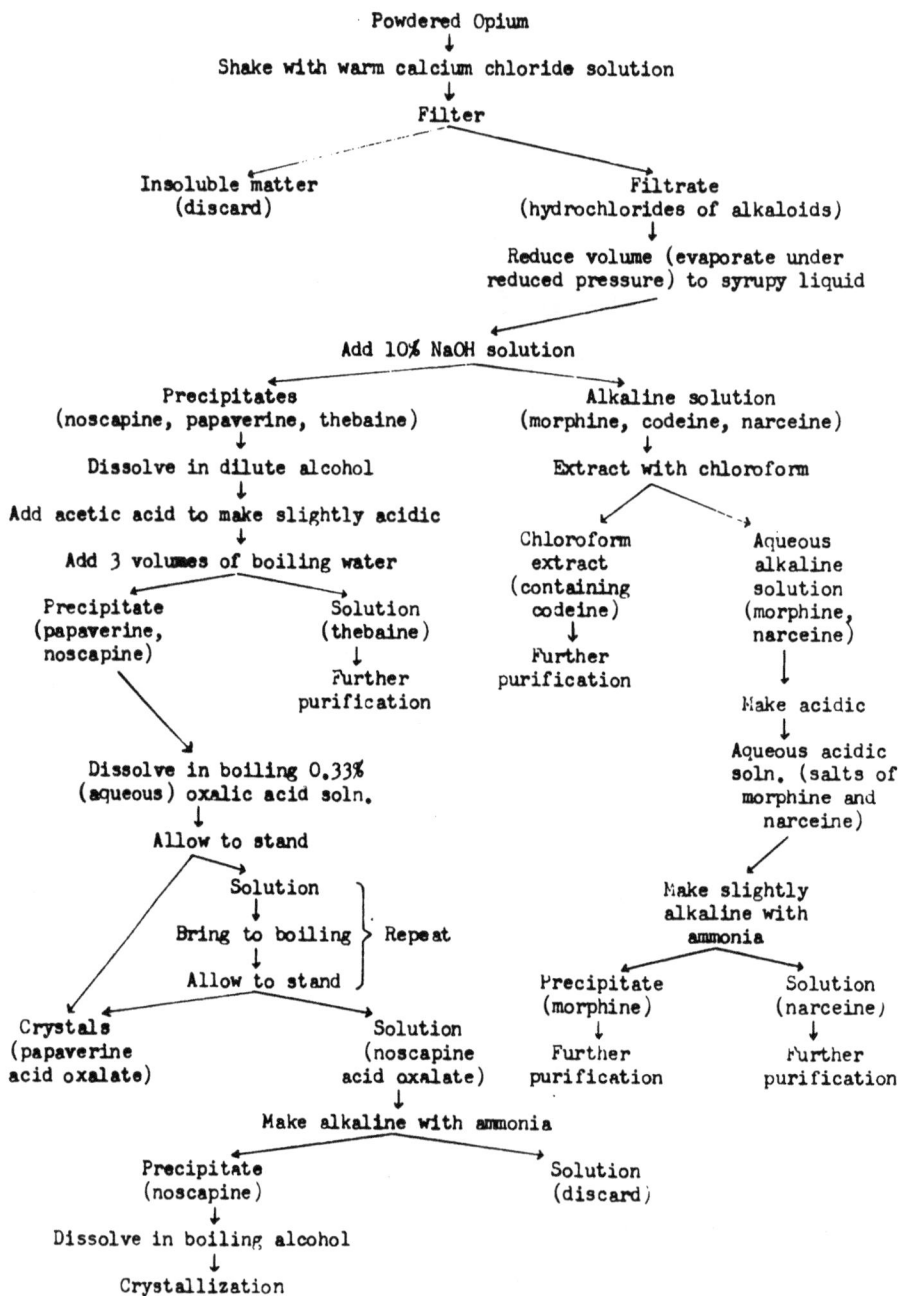

Figure 3. Procedure Outline for Isolation of Alkaloids of Opium

- 62 -

Tyrosine

Tyrosine

(−)-Norlaudanosoline

Papaverine

(−)-Reticuline

Salutaridine

Thebaine

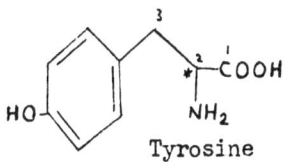

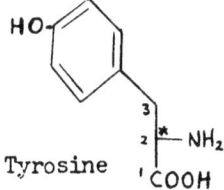

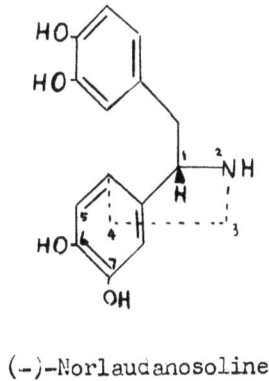

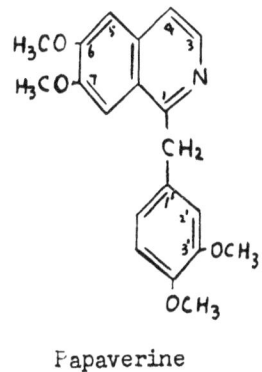

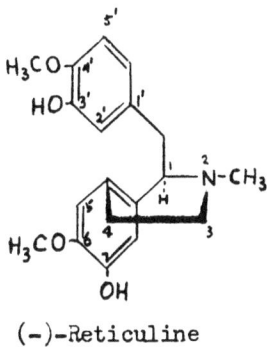

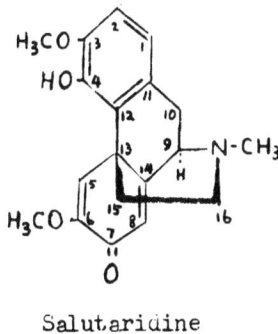

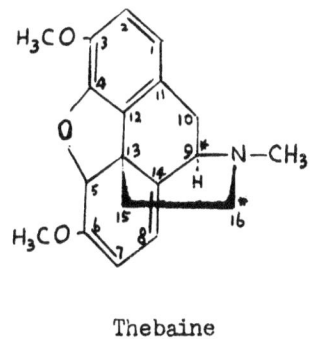

molecules (with label at C-9 and C-16), presumably also via nor-laudanoso-line (3, 10, 11). On the basis of very elegant experiments using quadruply-labelled and optically-active precursors, a plausible route from (−)-nor-laudanosoline through 1,2-dehydroreticuline, (−)-reticuline, salutaridine and salutaridinol to thebaine and other morphine alkaloids has been suggested (1). The obligatory role of 1,2-dehydroreticuline remains to be established.

(e) Pharmacological Action and Uses

The pharmacological action (13) of opium is mainly due to its mor-phine content. Morphine exerts a depressant action on the central nervous system, involving the cerebral cortex, hypothalamus, and medullary centres (narcotic action). This depression is especially marked on the perception of pain (analgesic effect), and on respiration. Morphine also exerts a stimulating effect on the spinal cord and on the vomiting centre. Morphine and its derivatives are important addicting drugs.

Codeine resembles morphine in its action, but it is much weaker. There is less tendency to tolerance and addiction with codeine than there is with morphine. There is greater risk of addiction with the synthetic derivative dihydrocodeinone than with codeine.

Among the synthetic derivatives of morphine, dihydromorphinone (hydromorphone, dilaudid), methyldihydromorphinone (metopon), oxymorphine, and levorphanol are all more potent analgesics than morphine, but also have addicting properties. Dihydromorphinone is also more toxic than morphine. Heroin (diamorphine) is also more toxic and there is greater risk of addiction. The use of heroin is outlawed in the U.S.A. and Canada, but it is used in certain European countries as a therapeutic agent.

Dihydrocodeinone (hydrocodone) and dextromethorphan are used chiefly for the relief of cough.

Apomorphine has little depressant action on the central nervous system, but it is a very effective emetic and expectorant.

Ethylmorphine resembles codeine in its action. It is chiefly used to produce vasodilatation of the conjunctiva in inflammatory conditions of the eye.

The benzylisoquinoline alkaloids of opium have no narcotic properties, and they exert no analgesic effect. Narceine is not used as a medicinal agent. Papaverine relaxes smooth muscle (acting directly on the muscle)

and prolongs the refractory period of the heart. It has been used in peripheral thrombosis and embolism, in bronchial asthma and myocardial infarc-
tion, but its effectiveness has been questioned. Noscapine is used as an
antitussive agent. It also relaxes smooth muscles.

(f) Medicinal Preparations

The principal preparations in U.S.P. XVII, N.F. XII, B.P.C. 1963,
and B.P. 1963 are listed below:

Opium U.S.P., B.P.C.

Opium Tincture B.P.

Paregoric U.S.P. (Camphorated Opium Tincture B.P.)

Ipecac and Opium Powder (Dover's Powder) B.P.

Papaveretum B.P.C.

Apomorphine Hydrochloride B.P.C.

Codeine Phosphate U.S.P., B.P.C.

Codeine Sulphate N.F.

Ethylmorphine Hydrochloride N.F., B.P.C.

Hydrocodone Bitartrate N.F.

Hydromorphone Hydrochloride N.F.

Morphine Hydrochloride B.P.C.

Morphine Sulphate U.S.P., B.P.C.

Noscapine U.S.P., B.P.C.

Noscapine Hydrochloride N.F.

Papaverine Hydrochloride N.F., B.P.C.

Papaverine Sulphate B.P.C.

Examples of pharmaceutical specialties containing dihydrocodeinone or its salt are: Codone[R], Dicodid[R], and Mercodinone[R]. The preparation

Copavin[R] contains papaverine hydrochloride and codeine sulphate.

5.3 Alkaloids of Ipecac

The drug Ipecac consists of dried root and rhizome of <u>Cephaelis Ipe-
cacuanha</u> (Brotero) A. Richard (Brazilian Ipecac), or of <u>Cephaelis acuminata</u>
Karsten (known as Cartegena Ipecac or Panama Ipecac). These plants are tropical shrubs, belonging to the Rubiaceae family. The roots and rhizomes
contain 2 - 2.5% total alkaloids. The two major alkaloids in Ipecac are
emetine (non-phenolic) and cephaeline (phenolic). In addition there are a
number of other alkaloids occurring in smaller quantities in these plants.

Emetine contains, in its molecule, two tetrahydroisoquinoline nuclei.
Cephaeline differs from emetine in that one of the four methoxyl groups in
emetine is replaced by a phenolic hydroxyl group in cephaeline (assumed to
be at position 6) (7).

Emetine is a white amorphous powder. It is laevorotatory, readily
soluble in alcohol, ether, or chloroform, less soluble in benzene or petroleum ether, and sparingly soluble in water. Its hydrochloride salt,

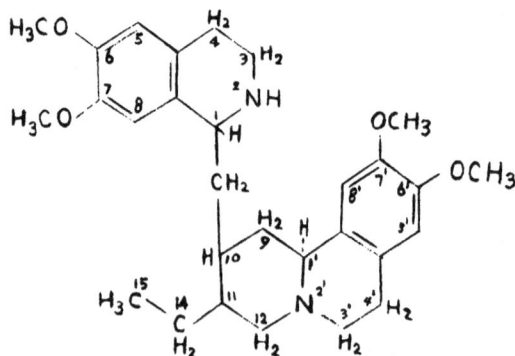

Emetine

(B).2HCl.7H$_2$O, separates from hot water as colourless needles, and the nitrate crystallizes from alcohol or water.

Cephaeline is readily soluble in chloroform or alcohol, less soluble in ether, insoluble in water, but soluble in aqueous solutions of alkalis. When cephaeline is treated with methyl sulphate, it yields emetine, N-methyl-emetine, and N-methylcephaeline.

In an aqueous acid extract obtained from a preliminary extraction of total alkaloids from Ipecac, emetine and cephaeline may be separated in the following manner: When the aqueous acid extract (containing the alkaloids in their salt forms) is treated with an excess of sodium hydroxide solution, the alkaloids are converted to their free-base forms, and emetine is precipitated (the free base being water-insoluble), while cephaeline, which has a phenolic hydroxyl group, remains dissolved in the strongly alkaline solution. The emetine may then be removed and further purified. To extract cephaeline from the solution, the latter is first neutralized, and then made weakly alkaline with ammonia or sodium bicarbonate and extracted with ether (5a).

Pharmacological Action and Uses

Emetine and cephaeline exert emetic action (from irritating effect on the stomach). Cephaeline is the more toxic of the two. In appropriate (smaller) doses, they produce expectorant effect. As expectorant in medicinal preparations, more often the crude drug Ipecac, rather than the pure alkaloids, is used, as Ipecac acts more slowly than the pure alkaloids and it has a lesser tendency to cause purging because of its tannin content. It is often used in combination with an antitussive agent. Ipecac has also been used as a diaphoretic by itself or in combination (as in Dover's Powder).

Emetine has been used in the treatment of amoebiasis (in amoebic dysentery), but this use is now largely superceded by other agents because the rather large doses required would cause excessive irritation of the stomach and emetic effect.

Medicinal Preparations in U.S.P. XVII, B.P. 1963, and B.P.C. 1963

      Ipecac U.S.P., B.P.C.

      Ipecac and Opium Powder B.P.

      Ipecac Liquid Extract B.P.C.

      Ipecac Syrup U.S.P.

      Ipecac Tincture B.P.

      Emetine Hydrochloride U.S.P., B.P.C.

5.4 Alkaloids of Curare

(a) Plant Sources

Curare is a crude dried plant-extract used as arrow poison by certain native tribes of the Amazon regions of South America. The combination of various plant species used in making this extract varies with different preparations of Curare, and often includes plants which apparently do not contain alkaloids with curariform pharmacological activity as well as plants which do. The identity of the plants which are used to make the Curare preparations exported from these South American regions is, in many cases, not well established (8, 14). The plant species which appear to be better documented in the literature as being the source materials from which various Curare preparations are made include Chondodendron tomentosum Ruiz et Pavon,

Chondodendron candicans (L. C. Rich.) Sandw., Chondodendron polyanthum Diels
(Fam. Menispermaceae), and Strychnos toxifera Bentham (Fam. Loganiaceae).
Other species of Chondodendron and of Strychnos have also been reported as
sources for Curare. The genus name Chondodendron is also spelled as Chon-
drodendron by some authors.

The extract, which constitutes Curare (as made by the natives), is
usually made by infusion of the bark and stems with hot water (200 gm. of
powdered plant material to about 2 litres of water) over a period of several
days, removal of the plant material by straining, and subsequent slow evaporation of the aqueous extract to a brown to black syrupy mass (8).
Various preparations of Curare have been called "Calabash" (gourd) Curare,
"Tube" (bamboo) Curare, and "Pot" (clay pot) Curare according to the con-
tainers used in packaging them.

(b) Tubocurarine Chloride and Other Curare Alkaloids

Among several alkaloids or alkaloidal salts which have been isolated
from Curare, (+)-tubocurarine chloride (d-tubocurarine chloride) is the only
one that is of pharmacological and therapeutic importance at present.
Several chemically synthesized derivatives are also useful medicinal agents.

(+)-Tubocurarine chloride was first isolated in crystalline form in
1935 by King (9). It is a quaternary bis-benzyltetrahydroisoquinoline base.
It is soluble in water or alcohol, and insoluble in ether or chloroform. On
treatment with methyl iodide in the presence of methanolic potassium hydroxide, O-dimethyltubocurarine iodide is obtained. This derivative and dimethyltubocurarine chloride are more potent in pharmacological action than (+)-
tubocurarine chloride, and both are used for the same therapeutic purposes

as the naturally occurring compound.

Occurring in Curare are also a number of alkaloids which are closely related to (+)-tubocurarine in chemical structure but which are tertiary bases and not quaternary bases. Among these is curine, which is (-)-bebeerine (1-bebeerine). Curine differs from (+)-tubocurarine in that the two nitrogens are in tertiary and not quaternary form, and that there is one (instead of two) methyl group attached to each of the two nitrogens. Bebeerine (the dextrorotatory isomer of curine) is also known as pelosine or chondodendrine (or chondrodendrine). Bebeerine and curine also occur in Bebeeru Bark, also known as Greenheart Bark (Nectandra Rodiaei Hook, Fam.

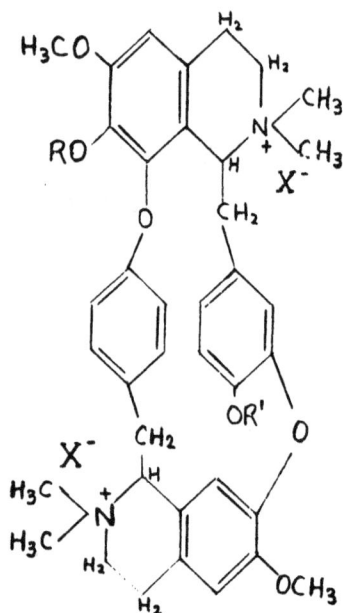

(+)-Tubocurarine Chloride: R = R' = H; X = Cl

Dimethyltubocurarine Iodide: R = R' = CH$_3$; X = I

Lauraceae), and in "Pareira brava" which is said to be the root of Chondodendron platyphyllum (St. Hil.) Miers, or of Chondodendron microphyllum
(Eichl.) Moldenke, or of Cissapelos Pareira L. (Fam. Menispermaceae). Other
plant species have also been reported to be the sources of "Pareira". Pareira has not been reported to contain (+)-tubocurarine.

(c) Isolation of (+)-Tubocurarine Chloride

King's procedure (9) for the isolation of curine and tubocurarine
chloride from stems of Chondodendron tomentosum may be summarized in outline
as follows: The powdered stem material (1.08 Kg.) is extracted with 15
litres of 1% tartaric acid solution, and the extract concentrated to 3.6
litres. The isolation of tubocurarine chloride and its separation from the
non-quaternary bases (curine, etc.) using 250 ml. of this tartaric acid extract may be represented by the schematic outline shown in Figure 4.

(d) Pharmacological Action and Medicinal Uses

Curare and (+)-tubocurarine prevent the effector cells or substance
at the myoneural junction of voluntary (skeletal) muscles from responding to
acetylcholine, and decrease the amplitude of the end-plate potential. They
prevent synaptic transmission of nerve impulses between the preganglionic
and postganglionic fibres of the autonomic nervous system. These effects
are counteracted by acetylcholine, physostigmine, and neostigmine.

The muscle-relaxant properties of (+)-tubocurarine are dependent on
the presence of, and the distance (13 - 15 Å) between the two quaternary
groups. Such structural considerations have led to the chemical synthesis
of other curarizing compounds.

Purified extract of Curare, (+)-tubocurarine chloride and its

Tartaric acid extract (250 ml.)
↓
Treat with 100 ml. of 0.5 N.
basic lead acetate solution
↓
Filter
↓
Filtrate
↓
Remove lead as PbS ppt. by
passing hydrogen sulphide
↓
Drive off H₂S from filtrate
↓
Make alkaline with sodium bicarbonate (25 gm.) and
stand for several days (a solid separates out)

Solid
↓
Extract with
chloroform
↓
Filter

Filtrate
(non-quaternary
bases: curine,
etc.)

Chloroform-
insoluble
solid
↓
Neutralize
with N HCl
(5 ml.)
↓
Treat with 40 ml.
of saturated soln.
of ammonium reineckate
↓
Precipitate
↓
Convert to chloride
↓
Crystalline
(+)-tubocurarine chloride
(0.32 gm.)

Aqueous solution
↓
Extract repeatedly
with chloroform

Aqueous soln.
↓
Make weakly
acidic with
conc. HCl
(27.5 ml.)
↓
Treat with 100 ml.
of saturated soln.
of ammonium reineckate
↓
Tubocurarine
reineckate
ppt.
↓
Convert to
chloride
↓
Concentration and
crystallization
↓
Crystalline
(+)-tubocurarine
chloride (0.37 gm.)

Chloroform ext.
(non-quaternary
bases: curine,
isochondoden-
drine)

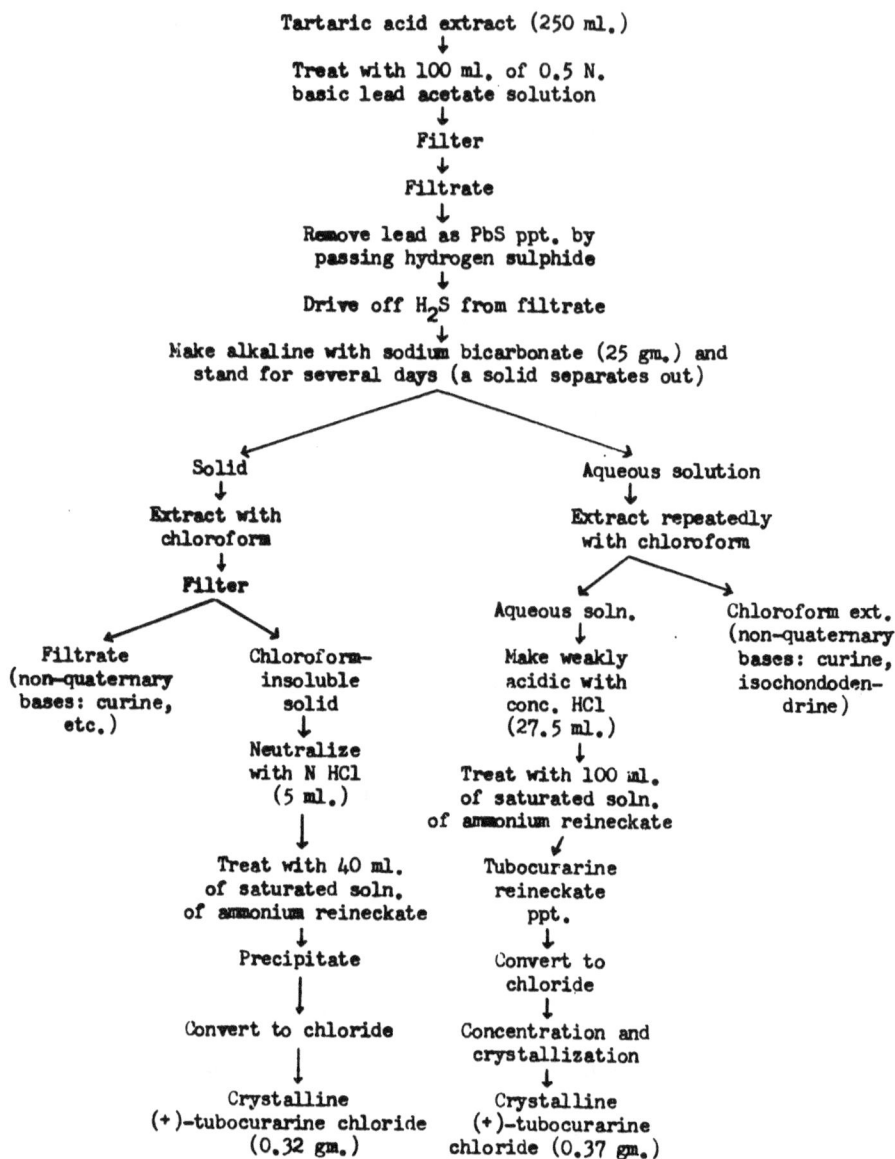

Figure 4. Procedure Outline for Isolation of Tubocurarine

dimethyl derivatives are used to induce greater muscular relaxation in light surgical anaesthesia. They are also used to reduce the severity of convulsion in shock treatments of mental disease and as a diagnostic aid in myasthenia gravis.

Bioassay

Curare preparations are assayed for potency on rabbits by the "head-drop crossover" test, in which a group of animals for testing and a group for controls are used on alternate days. The standard "head-drop" dose is the smallest amount of the drug capable of producing a "full head-drop" in 50% or more of the test group. This dose or unit is equivalent to 0.15 mg. of (+)-tubocurarine chloride pentahydrate in activity.

(e) Medicinal Preparations

      Tubocurarine Chloride U.S.P. XVII, B.P.C. 1963

      Dimethyl Tubocurarine Chloride N.N.D.

      Dimethyl Tubocurarine Iodide N.F. XII

      Chondodendron tomentosum Extract Purified N.N.D.

      Some examples of pharmaceutical specialties are: Tubarine[R];

Tubadil[R]; Intocostrin[R]; Metubine Iodide[R]; Mecostrin Chloride[R].

5.5 References

1. Barton, D. H. R., Kirby, G. W., Steglich, W., & Thomas, G. M. - J. Chem. Soc., pp. 2423-2434 (1965); Battersby, A. R., Dobson, T. A., & Ramuz, H. - ibid., pp. 2434-2438 (1965); ibid., pp. 3323-3332 (1965).

2. Battersby, A. R., & Harper, B. J. T. - Proc. Chem. Soc. (London), p. 152 (1959).

3. Battersby, A. R., & Harper, B. J. T. - Chem. and Ind., p. 364 (1958); J. Chem. Soc., pp. 3526, 3534 (1962).

4. Clarke, E. G. C. - J. Pharm. Pharmacol., 9:187-192 (1957).

5. Cromwell, B. T. - in K. Paech and M. V. Tracey (eds.): Modern Methods of Plant Analysis, Vol. IV, (a) p. 428, (b) p. 450-460, Springer, Berlin, 1955.

6. Farmilo, C. G., & Genest, K. - in C. P. Stewart, & A. Stolman (eds.): Toxicology: Mechanism and Methods of Analysis, Vol. II, pp. 229-242, Academic Press, New York, 1961.

7. Janot, M. M. - in R. H. F. Manske & H. L. Holmes (eds.): The Alkaloids, Vol. III, pp. 363-394, Academic Press, New York, 1953.

8. Karrer, P. - J. Pharm. Pharmacol., 8:161-184 (1956).

9. King, H. - J. Chem. Soc., p. 1381 (1935); p. 265, 1945 (1948); pp. 955, 3163 (1949).

10. Kleinschmidt, G., & Mothes, K. - Z. Naturforsch., 14b:52-56 (1959).

11. Leete, E. - Chem. and Ind., p. 977 (1958).

12. Martello, R., & Farnsworth, N. R. - Lloydia, 25:176-185 (1962).

13. Reynolds, A. K., & Randall, L. O. - Morphine & Allied Drugs, Chapters I - IV, University of Toronto Press, Toronto, 1957.

14. Wintersteiner, O. - in D. Bovet, F. Bovet-Nitti, & G. B. Merini-Bettolo (eds.): Curare and Curare-like Agents, p. 153, Elsevier, New York, 1959.

CHAPTER 6

ALKALOIDS OF THE QUINOLINE GROUP

6.1 Cinchona Alkaloids

(a) Plant Sources

The medicinally useful alkaloids of the quinoline group are exemplified by the alkaloids of Cinchona. Over twenty alkaloids have been isolated from the bark of Cinchona Calisaya Weddell, Cinchona Ledgeriana (Howard) Moens et Trimen (these are known in commerce as "Yellow Cinchona"), and Cinchona succirubra Pavon et Klotzsch ("Red Cinchona") (Fam. Rubiaceae), and their hybrids, as well as from Cuprea Bark (Remijia pedunculata Flueck., and Remijia purdieana Wedd., Fam. Rubiaceae).

(b) Properties and Tests

Among these Cinchona alkaloids, the primary ones of practical importance are: quinine, quinidine, cinchonidine, and cinchonine.

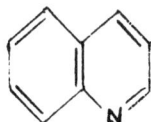

Quinoline

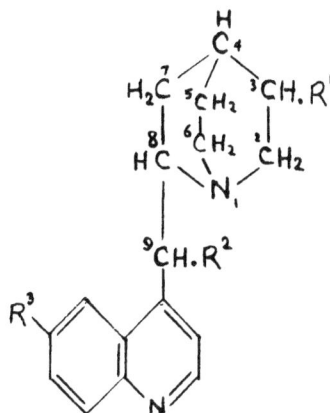

	$R^1=$	$R^2=$	$R^3=$
Ruban	H	H	H
Quinine, quinidine	$-CH = CH_2$	OH	$-OCH_3$
Cinchonine, cinchonidine	$-CH = CH_2$	OH	H

As may be seen from the structure formula, the parent structure (ruban) of these alkaloids contains a quinoline ring system attached through a methylene group to a dicyclic ring system.

Quinine (laevorotatory) and quinidine (dextrorotatory) are isomeric; so are cinchonidine (laevorotatory) and cinchonine (dextrorotatory).

By the numbering system (shown above) now commonly adopted, quinine and quinidine are, therefore, 6'-methoxy-3-vinylruban-9-ol, while cinchonine and cinchonidine are 3-vinylruban-9-ol.

The average quantities of these alkaloids in the dry bark materials from the above mentioned plant sources are: quinine 5 - 7%; quinidine 0.1 - 0.3%; cinchonidine 0.2 - 0.4%; cinchonine 0.2 - 0.4%. Other minor alkaloids in the bark are in smaller quantities than these.

Quinine

One gm. of anhydrous quinine dissolves, at 25° C., in 1750 ml. of water, 0.6 ml. of alcohol, 4.5 ml. of ether, 1.9 ml. of chloroform. It is readily soluble in boiling benzene, which is often used in extraction and isolation. Quinine crystallizes from boiling benzene in needles, containing solvent. The solvent is gradually lost on exposure to air. Quinine's solubility in ether is greater than that of the other Cinchona alkaloids, and this property is often utilized in separating it from the other alkaloids.

When quinine is precipitated by adding an aqueous solution of its acid sulphate to an excess of ammonia solution under agitation, the base may pass gradually from an amorphous form to a crystalline trihydrate form which changes to the dihydrate in air or to the monohydrate when dried over sulphuric acid, or becomes anhydrous when dried at 125° C. (5).

Quinine is a diacidic base, forming neutral and acid salts. The
sulphate is the most common salt form, and the alkaloid is usually isolated
from the plant material as sulphate. There are three forms of quinine sulphate: (a) a neutral salt, $(B)_2.H_2SO_4.8H_2O$ (or $7H_2O$), formed by neutralizing
the alkaloid (free base) with dilute sulphuric acid (5%), and recrystallized
from boiling water. It is only sparingly soluble in water (1 in 720 at 25^0
C.). This heptahydrate salt, when exposed to dry air, changes to the more
stable dihydrate, $(B)_2.H_2SO_4.2H_2O$. (b) The acid sulphate, $(B).H_2SO_4.7H_2O$
(quinine bisulphate), is soluble in water (1 in 8.5 at 25^0 C.) and in alcohol
(1 in 18). Its aqueous solution is acid to litmus. (c) The "tetrasulphate",
$(B).2H_2SO_4.7H_2O$ is very soluble in water.

Quinine hydrochloride, $(B).HCl.2H_2O$, resembles the neutral sulphate,
but is much more water-soluble (1 in 18 at 25^0 C.). When a solution of the
acid sulphate of quinine is treated with barium chloride, the acid hydrochloride of quinine, $(B)2HCl$, is formed.

Quinine also has other salt forms, such as the hydrobromide, the di-
hydrobromide (acid hydrobromide), the oxalate, etc. Quinine also forms addition compounds with a variety of organic compounds (e.g., sulphanilamide,
pantothenic acid, etc.) and mercury compounds.

The specific rotation of quinine varies widely with temperature,
solvent, and pH.

When quinine is heated with potassium hydroxide in amyl alcohol it
is isomerized to quinidine among other products (epimers of quinine and of
quinidine).

Quinidine

Quinidine (the dextrorotatory isomer of quinine) crystallizes from boiling alcohol with 2.5 mol. of solvent, from ether with 1/3 mol. of solvent, and from boiling water with 1.5 H_2O. One gm. of quinidine dissolves in 26 ml. of 80% alcohol at 20° C., in 35 ml. of ether at 10° C. Its solubility in ether, next to that of quinine, is also greater than that of the other Cinchona alkaloids, and this is also sometimes utilized in separation procedures. Quinidine is sparingly soluble in water (1 in 2000 at 15° C.), and in chloroform, and nearly insoluble in petroleum ether.

Quinidine is alkaline in solution, and it behaves as a diacidic base, forming neutral and acid salts. Its neutral sulphate, $(B)_2.H_2SO_4.2H_2O$, and the acid sulphate, $(B).H_2SO_4.4H_2O$, are crystalline, and are soluble in water or alcohol.

The neutral hydriodide of quinidine $(B).HI$, only very sparingly soluble in water (1 in 1250 at 15° C.), is formed as a crystalline powder when potassium iodide is added to a neutral aqueous solution of a quinidine salt. Quinidine hydriodide is much less soluble than that of the other Cinchona alkaloids, and quinidine is usually isolated in this form.

(c) Colour Reactions and Tests

(i) <u>Thalleioquin Test</u>: - When a few drops of bromine water are added to 2 or 3 ml. of a weakly acidic solution of a quinine salt, followed by the addition of 0.5 - 1 ml. of strong ammonia, a characteristic emerald green colour is produced. The coloured product has been called thalleioquin for which the chemical constitution is not known. This reaction may be given by quinine in concentration as low as 0.005 per cent. This reaction is also given by quinidine and by another Remijia alkaloid, cupreine, but not by cinchonine or cinchonidine.

(ii) Ferrocyanide Test: - In this test, a small quantity (10 - 15 mg.) of a quinidine salt is well mixed with 0.5 ml. of bromine water in an evaporating dish, and transferred to a test tube with the aid of 1 ml. of water. One ml. of chloroform is then added, and the solution is allowed to stand for a few minutes. A drop of a 10% solution of potassium ferrocyanide and 3 ml. of a 5 N sodium hydroxide solution are added with shaking or stirring. The chloroform layer assumes a red colour. With a quinine salt given this test, the chloroform layer remains colourless after addition of the ferrocyanide and sodium hydroxide (3).

(iii) Quinine, quinidine, cinchonine, and cinchonidine give precipitates with many of the alkaloid reagents such as Kraut's or Dragendorff's, Hager's, and Mayer's reagents (see pages 9 and 10).

Cinchonine and Cinchonidine

Cinchonine and cinchonidine are also diacidic bases. Both of these two alkaloids are considerably less soluble in ether than quinine and quinidine. This property is utilized in separating them from quinine and quinidine. Cinchonine and cinchonidine crystallize from alcohol.

(d) Extraction and Estimation of Total Alkaloids

The dry powdered Cinchona bark material (15 - 20 gm.) is first well mixed with about 30% of its weight of calcium hydroxide or calcium oxide and sufficient quantity of 5% sodium hydroxide solution to make a paste. Then it is allowed to stand for a few hours. The mass is then transferred to a Soxhlet apparatus and continuous extraction is carried out with benzene.

The benzene extract is then shaken with successive portions of 5% sulphuric acid, by which the alkaloid bases are converted to their water-soluble sulphate salts. The aqueous acid extract is then made alkaline with ammonia and shaken with portions of chloroform (to extract the alkaloids, in their free-base forms, liberated from their salts). The chloroform extract may then be reduced in volume by distillation, and quantitatively transferred to a tared beaker, and evaporated to a residue which is then dried at 100° C. to constant weight, and the weight determined (1, 2).

(e) Isolation of Individual Alkaloids

For isolation of the individual alkaloids, a larger quantity of the Cinchona bark material is subjected to the first part of the extraction procedure described above until the aqueous acid extract is obtained. A procedure as represented by Figure 5 may then be carried out for the separation and isolation of the four principal alkaloids (2).

(f) Pharmacological Action and Uses

Quinine acts on almost all forms of protoplasm with a transitory stimulation of its activity followed by depression and cessation (death) of activity. Its pharmacological effects on various tissues and organs of the animal and human bodies include a weak increase of contraction of the uterus, and a local anaesthetic effect (4).

Quinine and Cinchona preparations are principally used for their antimalarial action. The causative organisms (the various Plasmodium species) are apparently more susceptible than other protozoa to the action of quinine, particularly when the organism is in its amoeboid stage. The sulphate is the commonly used form for such therapeutic purposes.

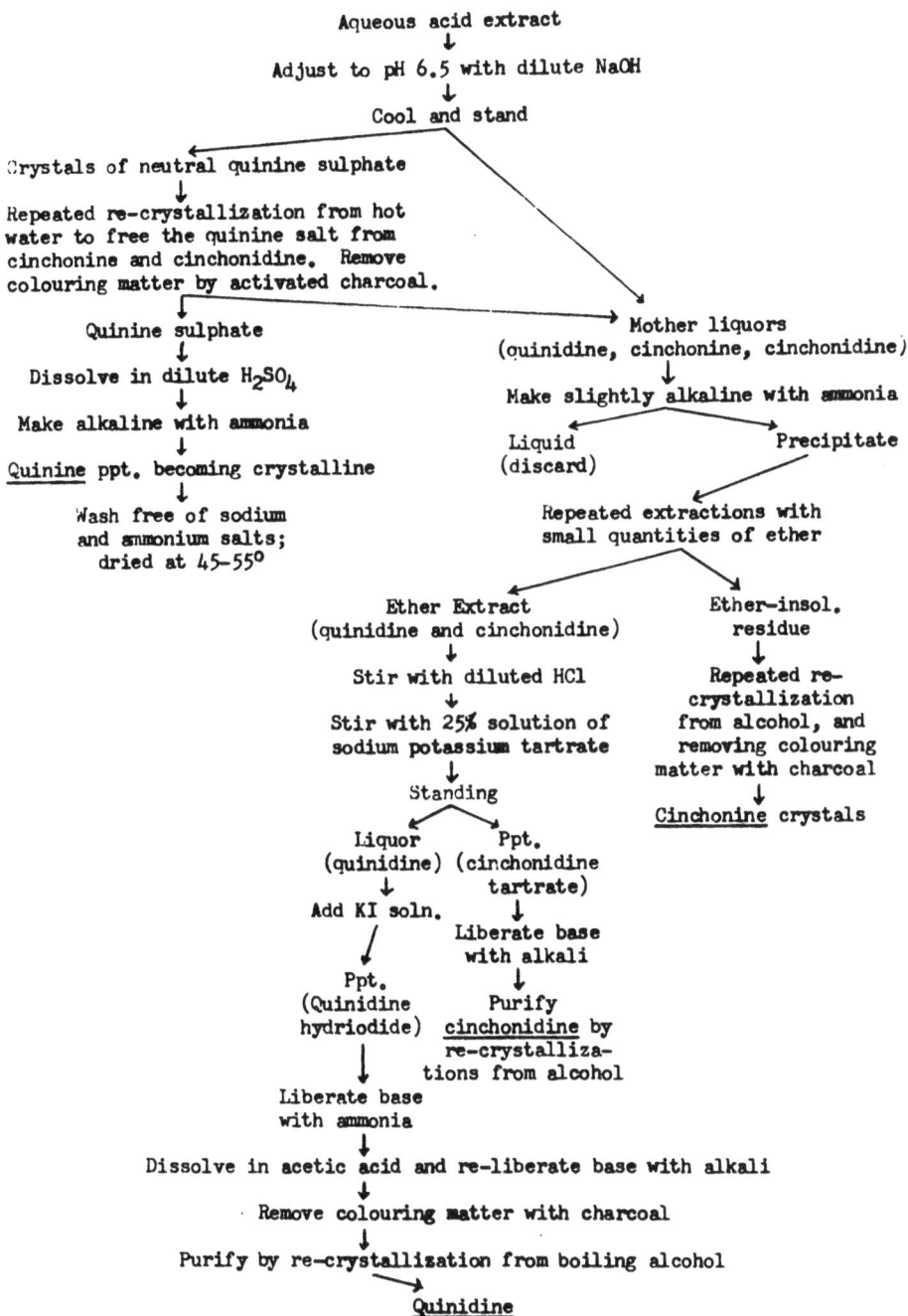

Aqueous acid extract
↓
Adjust to pH 6.5 with dilute NaOH
↓
Cool and stand

Crystals of neutral quinine sulphate
↓
Repeated re-crystallization from hot
water to free the quinine salt from
cinchonine and cinchonidine. Remove
colouring matter by activated charcoal.

Quinine sulphate
↓
Dissolve in dilute H_2SO_4
↓
Make alkaline with ammonia
↓
Quinine ppt. becoming crystalline
↓
Wash free of sodium
and ammonium salts;
dried at 45–55°

Mother liquors
(quinidine, cinchonine, cinchonidine)
↓
Make slightly alkaline with ammonia

Liquid   Precipitate
(discard)
↓
Repeated extractions with
small quantities of ether

Ether Extract   Ether-insol.
(quinidine and cinchonidine)   residue
↓   ↓
Stir with diluted HCl   Repeated re-
↓   crystallization
Stir with 25% solution of   from alcohol, and
sodium potassium tartrate   removing colouring
↓   matter with charcoal
Standing   ↓
                                 Cinchonine crystals
Liquor   Ppt.
(quinidine) (cinchonidine
↓   tartrate)
Add KI soln.   ↓
                Liberate base
                with alkali
Ppt.   ↓
(Quinidine   Purify
hydriodide)   cinchonidine by
                re-crystalliza-
↓   tions from alcohol
Liberate base
with ammonia
↓
Dissolve in acetic acid and re-liberate base with alkali
↓
Remove colouring matter with charcoal
↓
Purify by re-crystallization from boiling alcohol
↓
Quinidine

Figure 5. Isolation Scheme for Cinchona Alkaloids

- 81 -

Quinidine resembles quinine in its general action on protoplasm. It is principally used for restoring certain types of cardiac arrhythmia, such as auricular fibrillation and paroxysmal ventricular tachycardia. It depresses auricular and ventricular conductivity, prolongs the refractory period, and decreases the excitability of the myocardium (4).

(g) Medicinal Preparations

The principal preparations in U.S.P. XVII, N.F. XII, and B.P.C. 1963 are the following:

Cinchona B.P.C.

Quinidine Sulphate U.S.P., B.P.C.

Quinine B.P.C.

Quinine Bisulphate B.P.C.

Quinine Sulphate N.F., B.P.C.

Quinine Hydrochloride B.P.C.

6.2 References

1. Birch, H. F., and Doughty, L. R. - Biochem. J., 43:38-44 (1948).

2. Cromwell, B. T. - in K. Paech & M. V. Tracey (eds.): Modern Methods of Plant Analysis, Vol. IV, pp. 398-400, Springer, Berlin, 1955.

3. David, L. - Pharm. Acta Helv., 24:427 (1949).

4. Grollman, A. - Pharmacology and Therapeutics, 6th ed., pp. 509-511, 731-733, Lea and Febiger, Philadelphia, 1965.

5. Henry, T. A. - The Plant Alkaloids, 4th ed., p. 421, Blakiston, Philadelphia, 1949.

CHAPTER 7

ALKALOIDS OF THE QUINOLIZIDINE GROUP

7.1 Introduction

The quinolizidine (norlupinane) ring structure, in its actual or modified form, occurs in a number of alkaloids which collectively have been referred to in the literature as the "Lupin Alkaloids". These alkaloids have been isolated from a wide variety of plants, particularly in the Family Leguminosae, but also in certain other plant families. Among these quinolizidine alkaloids, the only one with medicinal significance is sparteine, which was first isolated more than one hundred years ago. Sparteine sulphate was official in the National Formulary VIII and in the British Pharmaceutical Codex 1949, but was omitted from subsequent editions of these compendia. It was formerly used in the treatment of certain cardiac arrhythmia. At present, there appears to be a renewed interest in its medicinal use, not as cardiac stimulant, but as an oxytocic agent.

7.2 Sparteine

(a) Plant Sources

Several alkaloids of the quinolizidine group were isolated, with empirical formulae correctly assigned to them, over a century ago. Structural elucidation for a number of these was achieved in the late 1920's and early 1930's. The first chemical synthesis of authentic quinolizidine from non-alkaloidal starting material was achieved in the early 1930's. The quinolizidine nucleus was unknown before its discovery, during this period, in the common "lupin alkaloids": lupinine, cytisine, sparteine, etc.

Sparteine occurs in some plants in the (-)-form, in other plants in the (+)-form, and still other plants as a mixture of the two forms. The more common form, (-)-sparteine (which is also called lupinidine), occurs in Chelidonium majus L., Cytisus proliferus L., Cytisus scoparius Link, Lupinus arboreus Sims, Lupinus barbiger S. Wats., and numerous other species (4). The dextrorotatory form, (+)-sparteine (which is also called pachycarpine), has been found to occur in Ammodendron conollyi Bge., Baptisia australis Lehm, Baptisia minor (L.) R. Br., Cytisus caucasicus Hort., Lupinus pusillus Pursh, Sophora pachycarpa C.A.Mey, and other species. (±)-Sparteine has been found to occur in Cytisus proliferus L. and other species. Of all the plant species listed here, Chelidonium majus is in the Family Papaveraceae, while all the other species are members of the Leguminosae.

Sparteine appears to occur in all parts of the plant, including the seeds. In Cytisus scoparius, which is known as the Common Broom, and Scotch Broom, the tops (stems and leaves) are usually used for extraction and isolation of the alkaloid, and the sparteine content in these parts of the plant has been found to vary from 0.3% to 1.5% in different months of the year (6). The seeds of many of these species also have a high sparteine content (2% in seeds of Lupinus arboreus), while the flowers generally contain very little alkaloid.

(b) Structure

(-)-Sparteine, also called lupinidine, was isolated and assigned its correct empirical formula $C_{15}H_{26}N_2$ in 1851, but its structural constitution was not fully elucidated until 1933. Its total synthesis was achieved in 1948 (4).

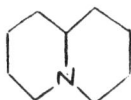

Quinolizidine

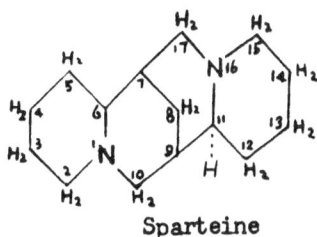

Sparteine

In the sparteine molecule, there are four asymmetric carbons at positions 6, 7, 9, and 11. The configurations of C-7 and C-9 are considered to be interdependent since the C-8 methylene bridge can only span the distance between C-7 and C-9 in a cis manner. Thus, it has been suggested (4) that the different configurations of stereoisomeric forms of sparteine may be conceived as arising from the relation of the hydrogen atoms on C-6 and C-11 to the C-8 methylene bridge, to give three racemic pairs: (a) hydrogens on C-6 and C-11 both cis to the C-8 bridge; (b) hydrogens on C-6 and C-11 both trans to the C-8 bridge; (c) hydrogen on C-6 cis, and hydrogen on C-11 trans.

(c) Properties

(-)-Sparteine is a liquid (b.p. 118° at 18 mm.), having a specific rotation (in ethanol) of -17°. It is sparingly soluble in water (1 in 328 at 22° C.), easily soluble in ethanol, chloroform and ether, and insoluble in benzene or petroleum ether. Sparteine sulphate, $(B).H_2SO_4.5H_2O$ forms rhombohedra crystals and is soluble in water and in ethanol. Its picrate crystallizes as yellow needles from boiling ethanol. It has pK_a of 4.54 and 11.78. Sparteine can be steam-distilled.

Oxidation of (±)-sparteine with potassium ferricyanide gives (±)-oxysparteine (i.e., with keto group at position 17 of the sparteine molecule) which cannot be reconverted to sparteine by catalytic reduction. Chromic acid-sulphuric acid oxidation of sparteine gives γ-aminobutyric acid.

Sparteine gives a white precipitate with Mayer's reagent and with silicotungstic acid. With Marme's reagent it forms crystals in plate and needle forms. The modified Grant's Test is sometimes used as a qualitative test for sparteine. This may be carried out by allowing a small quantity of a chloroform solution of sparteine to evaporate on filter paper. The residual spot on the paper is then exposed to bromine vapour for a few seconds when a yellow stain develops. The colour disappears on exposure to ammonia fume. If the spot is now warmed over a hot plate, a pink colour appears (2).

(d) Quantitative Estimation

In the procedure of White (6) for the quantitative estimation of sparteine, the powdered plant material is extracted in Soxhlet extractor with 2% acetic acid in 50% ethanol. The extract, after being concentrated to a smaller volume, is clarified with lead acetate, made alkaline and then distilled into dilute acid until the distillate gives no turbidity with potassium triiodide. The distillate is then concentrated to a small volume (5 ml.), and excess potassium cadmium iodide solution is added to precipitate the alkaloid (as its iodo-cadmium salt). After standing for 30 minutes, the precipitate is filtered through sintered glass filter, washed twice with dilute potassium iodide solution and once with water, dried at 40° C., and weighed as the alkaloid salt $(B).H_2CdI_4$.

(e) Pharmacological Action and Uses

Sparteine possesses, primarily, oxytocic as well as cardiac and respiratory stimulant actions (5). Its sulphate has been proposed for use to accelerate spontaneous labour or in elective induction of labour in term (in intramuscular dosage of 100 - 150 mg.). There have been several recent clinical studies of sparteine sulphate (1, 3), leading to the conclusion that it is a frequently effective but not an altogether safe oxytocic (3). Examples of pharmaceutical specialties containing sparteine sulphate are Sparto-cin[R], and Tocosamine[R].

7.3 References

1. Cromer, D. W., Reeves, B. D., & Danforth, D. N. - Am. J. Obstet, Gyne-col., 89:268-271 (1964).

2. Cromwell, B. T. - in K. Paech & M. V. Tracey (eds.): Modern Methods of Plant Analysis, Vol. IV, pp. 461-462, Springer, Berlin, 1955.

3. Filler, W. W., Jr., Filler, N. W., & Zinsberg, S. - Am. J. Obstet, Gyne-col., 88:737-747 (1964).

4. Leonard, N. J. - in R. H. F. Manske & H. L. Holmes (eds.): The Alkaloids, Vol. III, pp. 126, 138, 156-166, and Vol. VII, pp. 273-277, Academic Press, New York, 1953 and 1960.

5. Reynolds, A. K. - in R. H. F. Manske and H. L. Holmes (eds.): The Alka-loids, Vol. V, p. 180, Academic Press, New York, 1955.

6. White, E. P. - New Zealand J. Sci. Tech., B25:93-114 (1943).

CHAPTER 8

ALKALOIDS OF THE INDOLE GROUP

The important medicinal alkaloids which contain indole or indole

derivative in the structural nucleus consist of the alkaloids of Ergot,

Rauwolfia, Physostigma (Calabar bean), Nux Vomica, and the Vinca alkaloids.

8.1 Alkaloids of Ergot

(a) Sources

More than a score of alkaloids have been isolated from the fungus

Claviceps purpurea growing as parasite on various members of the Gramineae

(grasses) plant family. Certain other species of Claviceps also produce the

ergot alkaloids. Different strains of the Claviceps purpurea species iso-

lated from different host plants or grown in saprophytic cultures give rise

to different compositions of the ergot alkaloids produced.

The drug Ergot of the B.P.C. consists of the sclerotium of Claviceps

purpurea (Fries) Tulasne (Fam. Hypocreaceae, Class Ascomyceteae) developed

on rye plants, Secale cereale L. (Fam. Gramineae). But the ergot fungus is

capable of growing as parasite on many other members of the Gramineae family,

such as barley, oat, rice, wheat, etc. (11).

The ascospores or conidiospores of the fungus are carried by wind or

insects to lodge on the young ovaries of the host plants where they germin-

ate to produce hyphae. As the fungus grows parasitically on the host plant's

ovaries, it secretes a honey-like substance, called "honeydew", which

attracts insects by which the fungus spores are further disseminated. This

ends what is called the sphacelia stage. The mycelial threads of the fungus

penetrate deeper and deeper into the ovary and form a dense tissue finally
replacing the tissues of the ovary, forming a hardened purplish structure
called the sclerotium. This is the resting stage of the fungus' life cycle,
and this sclerotium contains the ergot alkaloids.

In nature, the sclerotium falls to the ground, and the following
spring it germinates into fruiting bodies (ascocarps) in each of which are
imbedded invaginations (called perithecia). Sacs (called asci) develop at
the base of the perithecium, and eight ascospores are formed in each ascus.
When the ascus ruptures, the spores in it are discharged and disseminated by
wind and insects to begin a new life cycle of the fungus.

(b) Peptide-type Ergot Alkaloids

From the ergot sclerotium there have been isolated six isomeric
pairs of "peptide type" of ergot alkaloids, each pair comprising a laevorotatory, pharmacologically active alkaloid and its dextrorotatory isomer
which is practically inactive pharmacologically. Total alkaloidal content
of the sclerotium varies widely from 0.025% to 0.4% or even higher. The
N.F. XI and B.P.C. 1963 specify an alkaloid content of 0.15% or higher.

A strain of Claviceps paspali Stevens and Hall, isolated by Arcamone
et al. (6) from host plant Paspalum distichum L. has been found to be able
to produce lysergic acid type of alkaloids in saprophytic culture.

Since 1948, many new alkaloids, of another series, which for convenience have been called the "clavine type" of ergot alkaloids have been isolated from various special strains of Claviceps purpurea and certain other
Claviceps growing on different host plants and/or in saprophytic cultures
(50a). These clavine-type of ergot alkaloids on alkaline hydrolysis do not

yield lysergic acid or isolysergic acid. The peptide-type of ergot alkaloids on alkaline hydrolysis give (+)-lysergic acid (from the (-)-isomer of each pair) or (+)-isolysergic acid (from the (+)-isomers of each pair), among the products.

In the ergot sclerotium, in addition to these alkaloids, there also occur a number of bases and amino acids. Those which have been reported include tyramine, betaine, histamine, putrescine, tyrosine, histidine, tryptophan, leucine, aspartic acid, and ergothioneine. Ergot also contains ergosterol and pigments.

The six isomeric pairs of the peptide-type of ergot alkaloids which have been isolated and characterized are the following:

Laevorotatory	Dextrorotatory
Ergonovine (ergometrine)	Ergometrinine
Ergosine	Ergosinine
Ergotamine	Ergotaminine
Ergocristine	Ergocristinine
Ergocryptine	Ergocryptinine
Ergocornine	Ergocorninine

Ergonovine is also known as ergometrine and ergobasine. Ergocristine, ergocryptine, and ergocornine are constituents of the alkaloidal fraction known as ergotoxine which was formerly thought to be a single alkaloid. What was formerly known as ergotinine consisted predominantly of what is now known as ergocristinine (39a).

In 1964, Schlientz et al. (51) isolated from natural ergot a new alkaloid ergostine and established its chemical constitution. Ergostine occurs in very small quantities (0.5 - 1% of the total alkaloids). In its chemical structure, it falls between ergotamine and ergocristine: that is, in ergos-

tine the R group in the general structure formula (p. 92) is $-CH_2.CH_3$,
otherwise the structure is the same as ergotamine. It yields lysergic
acid on alkaline hydrolysis. Its corresponding stereoisomer is named
ergostinine.

Lysergic acid differs from isolysergic acid only in the configuration at C_8, that is, the spatial position of the carboxyl group (C_{17}) in
relation to C_8. In the (-)-series (1-series) of these six pairs of alkaloids (ergometrine, ergotamine, etc.), the spatial relation of C_{17} and C_8
is the same as that in isolysergic acid. Other than this one difference,
in each isomeric pair the (-)-isomer has the same chemical structure as the
(+)-isomer.

Hydrolysis

On alkaline hydrolysis by a combination of several methods, these
six pairs of ergot alkaloids give the products listed in Table 4. Any one
particular method of hydrolysis does not necessarily produce all the products
exactly as listed in Table 4 for a particular alkaloid. This will become
clear in the summary of the different methods of hydrolysis to be described
presently.

On alkaline hydrolysis, ergometrine (ergonovine) yields lysergic
acid and (+)-2-aminopropanol. The products of hydrolysis for the other five
pairs of alkaloids become easily understandable if we view each alkaloid's
structure as comprising five portions, namely, the lysergyl portion (up to
and including C_{17}), the amide nitrogen that is linked to C_{17}, and the three
component parts marked as a, b, and c enclosed by dotted lines in the structure formula given on page 92. For convenience we shall use lysergic acid

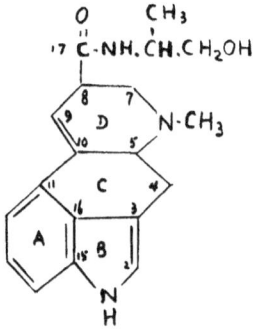

Ergonovine

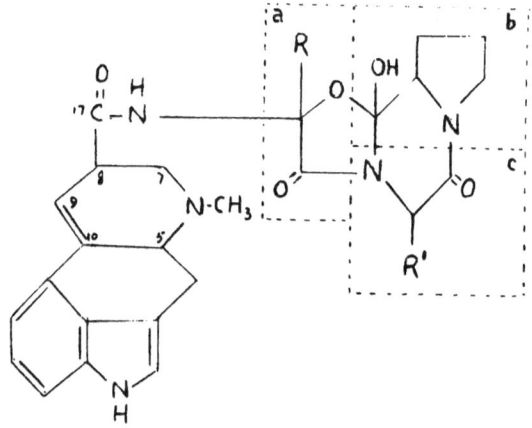

Ergot alkaloids containing
polypeptide side-chain

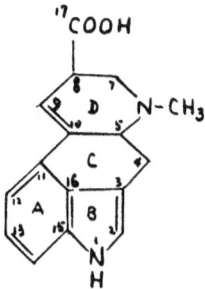

Lysergic
acid

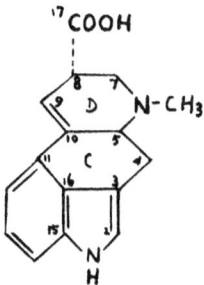

Isolysergic
acid

	R	R'
Ergosine	$-CH_3$	$-CH_2 \cdot CH \begin{smallmatrix} CH_3 \\ CH_3 \end{smallmatrix}$
Ergotamine	$-CH_3$	$-CH_2-\bigcirc$
Ergocristine	$-CH \begin{smallmatrix} CH_3 \\ CH_3 \end{smallmatrix}$	$-CH_2-\bigcirc$
Ergocryptine	$-CH \begin{smallmatrix} CH_3 \\ CH_3 \end{smallmatrix}$	$-CH_2 \cdot CH \begin{smallmatrix} CH_3 \\ CH_3 \end{smallmatrix}$
Ergocornine	$-CH \begin{smallmatrix} CH_3 \\ CH_3 \end{smallmatrix}$	$-CH \begin{smallmatrix} CH_3 \\ CH_3 \end{smallmatrix}$
Ergostine	$-CH_2 \cdot CH_3$	$-CH_2-\bigcirc$

Table 5. Products of Hydrolysis of Ergot Alkaloids

ALKALOID	Lysergic Acid	Iso-lysergic Acid	Proline	NH$_3$	Pyruvic Acid	Leucine	Phenyl-alanine	Valine	2-Amino-propanol	Dimethyl-Pyruvic Acid
Ergometrine	+	−	−	−	−	−	−	−	+	−
Ergometrinine	−	+	−	−	−	−	−	−	+	−
Ergosine	+	−	+	+	+	+	−	−	−	−
Ergosinine	−	+	+	+	+	+	−	−	−	−
Ergotamine	+	−	+	+	+	−	+	−	−	−
Ergotaminine	−	+	+	+	+	−	+	−	−	−
Ergocristine	+	−	+	+	−	−	+	−	−	+
Ergocristinine	−	+	+	+	−	−	+	−	−	+
Ergocryptine	+	−	+	+	−	+	−	−	−	+
Ergocryptinine	−	+	+	+	−	+	−	−	−	+
Ergocornine	+	−	+	+	−	−	−	+	−	+
Ergocorninine	−	+	+	+	−	−	−	+	−	+

in the following discussion of hydrolysis of these five pairs of alkaloids, but it is to be understood that the (+)-isomer in each pair contains an iso-lysergyl and not lysergyl moiety.

Hydrolysis by alcoholic alkali yields lysergic acid amide (lyserga-mide, ergine) from which ammonia is liberated by aqueous alkali. Hydrolysis of the alkaloid by aqueous alkali gives ammonia, lysergic acid, and an α-keto acid (derived from portion a), and the piperazide (comprising portions b and c). The α-keto acid so obtained is pyruvic acid in the case of ergo-sine or ergotamine, or their (+)-isomers, as the R in the formula (page 92) is a methyl group for these two pairs. The α-keto acid so obtained is di-methylpyruvic acid in the case of ergocristine-ergocristinine, ergocryptine-ergocryptinine, and ergocornine-ergocorninine, since the R in these three pairs is an isopropyl group. Hydrolysis of the alkaloid by hydrochloric acid destroys the lysergic acid but liberates the two amino acids derived from portions b and c (see structure formula on page 92). The amino acid so obtained from the portion b is proline. The amino acid so obtained from the portion c in each pair of these alkaloids is given in Table 6 in relation to the structure of the R' group in the general formula.

Degradation of the alkaloid with hydrazine splits off the lysergic acid as the hydrazide and leaves the whole polypeptide side-chain intact (comprising the three components a, b, and c shown in the general formula given on page 92) with the component a in the reduced form (fatty acid form) and not in the α-keto acid form. For example, with ergotamine, the poly-peptide so obtained is propionyl-L-phenylalanyl-L-proline. But with careful-ly controlled partial hydrolysis of the alkaloid with one equivalent of

Table 6. Amino Acids Obtained by Acid Hydrolysis of
Ergot Alkaloids Having Polypeptide Side Chain

Alkaloid	Amino Acids from Hydrolysis		R' group in the alkaloid
	From portion b	From portion c	
Ergosine-ergosinine Ergocryptine-ergocryptinine	Proline $O=C$—$C(OH)$—ring with N—H	Leucine H_2N, $COOH$ CH $CH_2 \cdot CH$ with Me, Me	$-CH_2 \cdot CH$ with Me, Me
Ergotamine-ergotaminine Ergocristine-ergocristinine	Proline	Phenylalanine H_2N, $COOH$ CH CH_2—phenyl	$-CH_2$—phenyl
Ergocornine-ergocorninine	Proline	Valine H_2N, $COOH$ CH CH Me Me	CH Me Me

aqueous alcoholic potassium hydroxide, the polypeptide so obtained contains
the acid in the α-keto-acid form (that is, for example, as pyruvoyl-L-
phenylalanyl-L-proline (39a, 50a, 54).

In the above discussion, we have viewed the products of hydrolysis
in relation to the different component parts of the alkaloid molecular
structure. In the history of the chemical investigations of these alkaloids

over the past three decades or so, the products of hydrolysis by the various methods as well as other reactions have served as the experimental evidence and basis on which the structures of these alkaloids have been deduced (39a, 50a, 54).

Reductive cleavage of ergocristine with sodium and butanol gives dihydrolysergol, ammonia, the acid (in the reduced form) derived from component 'a' (α-hydroxyisovaleric acid), and a dipeptide (comprising the 'b' and 'c' components) which can be hydrolyzed by acid to the two amino acids (derived from components 'b' and 'c' respectively).

These pairs of ergot alkaloids and lysergic acid undergo reduction of the C_9-C_{10} double bond by catalytic hydrogenation (e.g., over palladium black), to yield the dihydro alkaloids and dihydro lysergic acid. Dihydro-ergotamine, dihydroergocristine, dihydroergocryptine, and dihydroergocornine are also used as pharmacological and therapeutic agents.

Isomerization

The laevorotatory ergot alkaloids are converted to their corresponding dextrorotatory isomers by boiling their solutions in ethanol or methanol or by the action of alcoholic alkali. Ergotamine on standing in alcoholic solution is slowly converted to its (+)-isomer, ergotaminine. Ergometrine (ergonovine) in solution undergoes mutarotation.

Since the (+)-isomers are practically inactive pharmacologically, if conversion of the (-)-alkaloids to the corresponding (+)-isomers should occur in a medicinal preparation in storage, the pharmacological potency of that preparation would be correspondingly decreased.

These ergot alkaloids in aqueous acid solution on excessive exposure to light or on irradiation with ultraviolet light are converted to their "lumi" products. These "lumi-alkaloids" (lumi-ergotamine, etc.) differ from the corresponding parent alkaloids in having the C_9-C_{10} double bond of the lysergic acid component reduced to a single bond and having a hydroxyl group attached to C_{10} (39a). Therefore, the ergot alkaloids in storage, especially when in solution, need to be protected from excessive exposure to light if deterioration is not to occur.

All these ergot alkaloids show fluorescence under ultraviolet light. This property is often utilized in detecting the alkaloidal spots on developed paper chromatograms. Their lumi derivatives do not show such fluorescence under ultraviolet light. The lumi derivatives also exhibit the indole-type of ultraviolet spectrum with maxima at 224 and 285 mμ., whereas the ergot alkaloids show their characteristic maximum at or near 318 mμ. in their ultraviolet spectrum (39a).

(c) Colour Reactions and Tests

Van Urk Reagent (Ehrlich reagent): - With p-dimethylaminobenzaldehyde, the ergot alkaloids produce a characteristic deep blue colour. The form of the reagent (van Urk reagent) commonly used for this purpose consists of a 0.125% or 0.2% solution of p-dimethylaminobenzaldehyde in 65% sulphuric acid with addition of 0.2 ml. of 5% ferric chloride solution per 100 ml. of reagent. Some of the clavine-type of ergot alkaloids produce a greenish colour rather than this deep blue colour with this reagent. This reagent can also be adapted for use in spraying developed paper chromatograms of the ergot alkaloids, and for such purpose 10% hydrochloric acid is usually substituted for the sulphuric acid. One volume of a solution of the ergot alkaloids in concentrations as low as a few micro-grams per ml. when mixed with two volumes of the van Urk reagent is sufficient to produce the colour. Under standardized conditions quantitative estimation of total alkaloids in purified aqueous acid extracts derived from powdered ergot samples may be carried out by this colour reaction with the van Urk reagent when the intensity of the blue colour so produced is measured for absorbance (optical density) at 590 mμ. in a spectrophotometer and compared or matched with that produced by a standard solution of known concentration of a pure ergot alkaloid treated in the same way. The assay method for total alkaloids in ergot in the British Pharmaceutical Codex is based on this. Lysergic acid itself and certain indole derivatives when treated with this reagent also produce colours which are rather similar to the colour produced by the ergot alkaloids, usually more of a pinkish-blue rather than a deep blue colour.

Keller Test: - A solution of an ergot alkaloid in acetic acid containing a trace of ferric chloride shows a cornflower blue colour. This is known as the Keller Test. The ergot alkaloids as well as lysergic acid give the blue colour of this Keller test.

Among these ergot alkaloids, ergonovine (ergometrine) is water soluble, while all the others are nearly insoluble in water. They are all insoluble in petroleum ether. Petroleum ether is often used for removing oils from powdered ergot before extraction of the ergot for alkaloids is begun.

(d) Extraction and Estimation of Total Alkaloids

Powdered ergot, after being defatted by petroleum ether, is moistened with ammonia and then extracted with chloroform (or ether), and the extract so obtained is shaken with 0.2 N. sulphuric acid or 1-2% tartaric acid solution. The resulted aqueous acid solution, containing the alkaloid salts, is

then made up to an exact volume, and suitable aliquots of it are taken and

mixed with two volumes of van Urk reagent and the blue colour developed is

measured for absorbance in a spectrophotometer at 590 mμ. and compared to

that given by a solution of a reference standard ergot alkaloid treated with

the van Urk reagent in the same way, and the quantity of total alkaloids in

the unknown sample calculated.

For the estimation of the water-soluble ergonovine, the water-

insoluble alkaloids are removed from the acid aqueous solution containing

the total alkaloids by making the solution alkaline with ammonia and extract-

ing the other alkaloids from the aqueous solution by carbon tetrachloride.

The aqueous solution (containing ergonovine) is saturated with sodium chlor-

ide and extracted with ether, and the resulting ether extract may then be re-

extracted with dilute sulphuric acid or tartaric acid solution. Aliquots of

the acid solution can then be estimated for quantity of the alkaloid (ergo-

novine) by reacting with van Urk reagent as before.

(e) Isolation of the Individual Alkaloids

The procedures for the isolation of ergonovine, ergotamine, and the

constituents of ergotoxine may be summarized in outline as follows:

Isolation of Ergonovine

Defatted powdered ergot is extracted with hot dilute sulphuric acid

and the acid extract is then treated with excess of baryta (barium sulphate)

and the barium removed with carbon dioxide and filtration. The filtrate is

then concentrated by evaporation under reduced pressure. The concentrated

solution is treated with alcohol, made alkaline, and extracted with chloroform. The resulted chloroform extract is extracted with dilute sulphuric acid. The acid solution is made alkaline with ammonia and saturated with sodium chloride and then extracted with ether. The solvent is removed from the ether extract, leaving the alkaloidal residue from which ergonovine may be crystallized from acetone (15a, 16).

Isolation of Ergotamine

The method of Stoll (15a, 53), may be summarized as follows: The defatted powdered ergot is thoroughly mixed with aluminum sulphate and water and then subjected to continuous extraction with hot benzene. After removal of the solvent, the residue obtained is stirred for several hours with a large volume of benzene made alkaline with ammonia gas. After filtration, the benzene extract is concentrated, under reduced pressure, to about one-fiftieth of the original volume and ergotamine crystallizes out. A further quantity of ergotamine may also be crystallized from the mother liquor by treatment with petroleum ether. The ergotamine crystals may be recrystallized from aqueous acetone (with solvents of crystallization).

Isolation of the Ergotoxine Group

In Barger's method (15a), powdered ergot is extracted with ethanol and the solvent is then removed from the alcoholic extract. The residue thus obtained is defatted by petroleum ether, and then dissolved in ethyl acetate. The ethyl acetate extract is then shaken with 1 - 2% citric acid solution. Sodium bromide is added to the acid solution to convert the alkaloid salts to their hydrobromides which are then precipitated.

The hydrobromide precipitates are made alkaline with dilute sodium hydroxide solution and extracted with ether. The ether here removes the

ergocristinine (formerly ergotinine) and some ergocristine, while ergocryptine and ergocornine and some ergocristine remain in the aqueous alkaline solution. The aqueous solution is now neutralized and made slightly
alkaline with sodium carbonate solution and extracted with ether. Ether is
removed from the ether extract by distillation and the residue thus obtained
is dissolved in 80% ethanol and a slight excess of a phosphoric acid solution
added and allowed to stand for several days when ergotoxine phosphate crystals separate out and may be recrystallized from boiling 90% ethanol (50 ml.
for each gm. of the ergotoxine phosphate).

The crystalline ergotoxine is dissolved in absolute ethanol or methanol, and added to absolute ethanol containing two equivalents of di-(p-toluyl)-L-tartaric acid per equivalent of base. The ergocristine salt, which
is nearly insoluble in absolute alcohol, crystallizes out. The ergocryptine
salt, sparingly soluble in absolute ethanol, or absolute methanol, is separated by fractional crystallization in the same way. After ergocristine and
ergocryptine are removed (as the di-(p-toluyl)-tartrate), water is added to
the methanolic (or ethanolic) mother liquor until the methanol content is
80% when ergocornine crystallizes out and may re-crystallize from alcohol or
acetone (15a, 39a).

(f) Pharmacological Action and Uses of the Ergot Alkaloids

Ergonovine differs from the other ergot alkaloids in its ability to
cause prompt and vigorous contractions of the uterus in the puerpural period,
and hence its use (usually as the tartrate or maleate) for preventing hemorr-
hage in child-birth as administered after delivery of the placenta. Its
derivative, methyl-ergonovine, is used for the same purpose.

Ergonovine has little or no sympatholytic or adrenolytic action, while the alkaloids of the ergotoxine group possess this adrenolytic property. The ergotoxine group produce a reversal of the pressor effect of administered epinephrine. Ergotamine and the ergotoxine group constrict peripheral blood vessels to cause a rise in blood pressure and also inhibit the epinephrine contraction of isolated rabbit uterus. This latter property is used in the Bloom–Clark bioassay method for the ergot alkaloids. Ergonovine does not possess this property, but shares with the ergotoxine group and ergotamine the constricting action on peripheral blood vessels. The bioassay method using the cyanosis of the cock's comb (by the action of the ergot alkaloids) as the indicator, as was formerly used as a bioassay method, was based on this action. Ergotamine (usually as the tartrate) and dihydroergotamine are used for the treatment of migraine (by vasoconstriction of cerebral blood vessels).

The dihydro derivatives of the ergotoxine group of alkaloids possess vagotonic activity and greater sympatholytic and adrenolytic action than their corresponding unmodified parent alkaloids. They have only slight direct stimulating action on smooth muscle of the peripheral blood vessels and of the uterus.

The lysergic acid derivative lysergic acid diethylamide (LSD), which is not naturally occurring in ergot, also has ecbolic action, but it is not used therapeutically for this purpose. It produces a transient psychosis which sometimes resembles schizophrenia in some respects. Psychotic manifestations can result from doses as little as 0.3 mg. or less. It also produces a wide range of hallucinogenic and psychedelic effects which are affected in kind and in degree by the subject's personality, age, education, experience and other factors (30). There does not appear to be any general agreement among medical authorities regarding its usefulness as an aid to psychotherapy of certain types of psychiatric disorders or as a psychedelic agent in the treatment of alcoholics (30). Its clinical use for these purposes at present is restricted to studies by a limited number of specialists (30).

(g) Medicinal Preparations

Ergot B.P.C. 1963

Ergonovine Maleate U.S.P. (Ergometrine Maleate B.P.C.)

Ergotamine Tartrate U.S.P., B.P.C.

Methylergonovine Maleate U.S.P.

Methylergometrine Maleate B.P.C.

Examples of Pharmaceutical Specialties

Bellergal[R]; Cafergot[R]; Ergotrate[R]; Ergaloid[R]; Gynergen[R]; Hydergine[R]; Megral[R]; Methergine[R]; Neogynergen[R]; Wigraine[R].

(h) Clavine Type of Ergot Alkaloids

Since the isolation of agroclavine by Abe in 1948 (1), a number of other clavine-type of ergot alkaloids have been isolated principally in the laboratories of Abe (3, 4) and Stoll and Hofmann (31, 55) from various Claviceps strains parasitic on a variety of host plants such as Agropyron semicostatum Nees, Festuca rubra L., Elymus mollis Tri., Pennisetum typhoideum Rich., etc. Some of these clavine-type ergot alkaloids have also been found in saprophytic cultures of a number of these Claviceps strains and in Aspergillus fumigatus Fres. (2, 31, 52).

These clavine alkaloids are water-soluble ergoline derivatives, which do not yield lysergic or isolysergic acid on alkaline hydrolysis. Listed in Table 7 are the better known ones among the clavine-type of ergot alkaloids for which the chemical constitution has been fairly well established.

These clavine alkaloids also produce colour with the van Urk and the Keller reagents, but in many cases the colour so produced is not entirely the same as the deep blue produced by the peptide-type of ergot alkaloids, but rather as violet-blue, purplish-blue, or greenish or brownish in some cases. Setoclavine and penniclavine (and their stereoisomers), containing the C_9-C_{10} double bond conjugated with the indole ring system, show fluorescence under ultraviolet light, as the peptide-type ergot alkaloids do, while the other clavine alkaloids listed below do not.

Agroclavine on reduction with sodium and butanol yields festuclavine, pyroclavine, and costaclavine. Festuclavine and pyroclavine are stereoisomers, differing in configuration at C_8. Festuclavine and costaclavine are stereoisomers, differing in the configuration at C_{10}. Dichromate oxidation of agroclavine gives setoclavine and isosetoclavine, which are stereoisomers differing from each other in the configuration at C_8.

Agroclavine shows very active stimulating action on the rabbit uterus and some of the other clavine alkaloids have also been tested and found to possess ecbolic property. Their pharmacology has not yet been thoroughly investigated, and they are not used as medicinal agents for therapeutic purposes at present.

Table 7. Clavine-type of Ergot Alkaloids

Alkaloid	Constitutional characters of Ring D	Nature of UV spectrum	Reference
Agroclavine	C_8-C_9 double bond	Maxima at 227, 284, 293 mμ. (like that of uncon-jugated indole deriva-tives)	2, 3, 55
Festuclavine	No double bond in Ring D	Rather like that of un-conjugated indole deri-vatives	2, 52
Pyroclavine	"	"	3
Costaclavine	"	"	3
Elymoclavine	C_8-C_9 double bond	"	3, 55
Chanoclavine (Secaclavine)	Tricyclic; no ring D	"	31, 52a
Isochanoclavine	"	"	52a
Setoclavine (Triseclavine)	C_9-C_{10} double bond (conjugated with indole ring system)	Rather like that of ly-sergic acid	4
Isosetoclavine	"	"	4
Penniclavine	"	" (maxima at 240, 315 mμ.)	3, 31, 55
Isopenniclavine	"	"	3

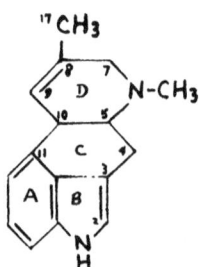

Agroclavine

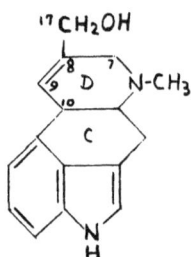

Elymoclavine

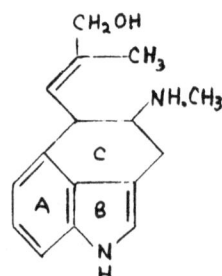

Chanoclavine

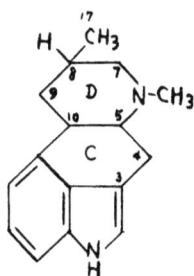

Festuclavine

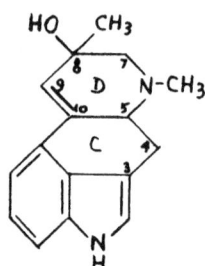

Setoclavine

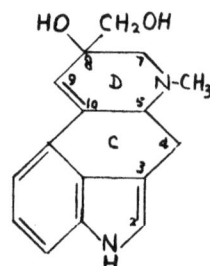

Penniclavine

A number of clavine-alkaloid-producing strains of ergot fungus are at present widely used in experimental studies on biosynthesis of the ergot alkaloids as they produce alkaloids fairly easily in saprophytic cultures, while those <u>Claviceps</u> strains which produce the peptide-type ergot alkaloids in parasitic growth are difficult to maintain in saprophytic culture for abundant alkaloid production.

(i) <u>Paper Chromatography</u>

A number of solvent systems have been devised for the paper chromatography of the ergot alkaloids. The following are four useful systems which have been used successfully by various authors:

(a) <u>Formamide-Benzene-Petroleum Ether System</u> - This system (22, 59) is useful for separating the water-insoluble peptide-type ergot alkaloids. Whatman No. 1 paper is dipped in formamide:absolute ethanol (1:1) and blotted before the alkaloids (in their base form and dissolved in chloroform:isopropyl alcohol 3:1) are applied. The chromatogram is developed by descending technique over a 10-12 hour period, using benzene:petroleum ether (b.p. 100-120°) 4:1 as the developing solvent.

(b) <u>Buffered System</u> - This system (12) has also been used for the separation of the peptide-type of ergot alkaloids. Whatman No. 1 filter paper is dipped in a 0.1 M citric acid-0.2 M disodium phosphate buffer solution, pH 3-4, blotted and air-dried. The alkaloids, in their base form, are dissolved in chloroform or 90% ethanol for spotting on to the paper, and the chromatogram is developed over 3-5 hours, using anaesthetic ether as the mobile phase.

(c) <u>Formamide-Benzene-Pyridine System</u> - This is a useful system which is frequently used for the separation of the water-soluble clavine-type ergot alkaloids (5, 47). The alkaloids in their base form may be dissolved in methanol or in chloroform:propanol (3:1) for spotting on Whatman No. 3 filter paper which has been previously dipped in a formamide:methanol (1:3) mixture containing 4% benzoic acid. A benzene-pyridine (6:1) mixture is used for equilibrating the chamber and as the developing solvent. The development of the chromatogram is allowed to proceed over a 3- to 4-hour period if the solvent front is to be retained on a 20-inch long paper strip for determination of R_f values. For better separation of the individual alkaloids in a mixture, the chromatogram may be allowed to be developed over an 8-hour period, in which case the solvent front is over-run.

(d) Butanol-Acetic Acid-Water System - This system has been used to separate ergometrine from ergometrinine (22) and these alkaloids are applied as the tartrate or maleate on Whatman No. 1 paper. The upper phase (butanol saturated with water) from a mixture of butanol:glacial acetic acid:water (4:1:5) is used as the developing solvent, and the development of the chromatogram is carried out in a 12-18 hour period. The water-insoluble ergot alkaloids (ergotoxine group, ergotamine, etc.) are carried along by the solvent front in this system (22).

Separation of the clavine-type ergot alkaloids has also been carried out on Whatman No. 1 paper with a developing solvent mixture of butanol:acetic acid:water (4:1:1) by descending technique over a 14-hour period (5).

The positions of both the peptide-type and the clavine-type of the ergot alkaloids on the paper chromatogram may be detected by spraying with a 1% solution of p-dimethylaminobenzaldehyde in 10% hydrochloric acid. The peptide-type ergot alkaloids and some of the clavine alkaloids (e.g. setoclavine) may also be detected on the developed chromatogram by their fluorescence under ultraviolet light. As previously pointed out, a number of the clavine-type ergot alkaloids do not show this fluorescence under ultraviolet light. For the application of the alkaloids on the paper with all the above systems, generally 5-15 µg. quantities are suitable.

(j) Thin-Layer Chromatography

Several solvent systems have been successfully used for thin-layer chromatography of the clavine-type ergot alkaloids. A useful one is that using ethylacetate:ethanol:N,N-dimethylformamide (13:1:1) as the developing solvent (29, 61). A rapid and simple method has been devised (61) for preparing small-sized (1 in. x 4 in. or 3 in. x 5 in.) chromatoplates by dipping a pair of such plates, sandwich fashion, in a well-stirred suspension of 80 gm. of silica gel G in 200 ml. of chloroform:methanol (2:1), and withdrawing the plates slowly from the suspension, resulting in two chromatoplates which dry rapidly and are ready for use in two or three minutes. Such plates can be used with the solvent system mentioned above for the separation of a number of the clavine-type of ergot alkaloids. With plates of the size mentioned above, the chromatogram may be developed over a 25- to 35-minute period.

(k) Biosynthesis of the Ergot Alkaloids

In recent years, several studies have been made on the parasitic growth of Claviceps regarding the precursors for the biosynthesis of the ergot alkaloids (see (A) below). Numerous other experimental investigations on the pathway of biosynthesis of these ergot alkaloids have been made on artificial cultures of Claviceps (see (B), (C), and (D) below). The most significant results of these experimental studies may be summarized as represented by the following:

(A) Gröger and Mothes (26) injected tryptophan into rye ovaries which were infected artificially with ergot fungus, and found increased alkaloidal content as compared to those ergot-infected rye ovaries that were not so injected with tryptophan. Mothes et al. (40) also injected tryptophan-β-C^{14} in a similar manner, and found incorporation of the isotope into the lysergic acid portion of the alkaloids.

(B) Tryptophan-β-C^{14} was fed to artificial cultures of alkaloid-producing strains of Claviceps by Baxter, Kandel, and Okany (7), by Gröger

- 109 -

et al. (27), and by Taber and Vining (60), and it was found that the isotope was incorporated into the lysergic acid-type alkaloids (ergosine, etc.). The C^{14} isotope from tryptophan-C^{14}OOH was not incorporated into the ergot alkaloids by such cultures (7).

These experimental results show that tryptophan serves as a precursor for the appropriate portion of the nucleus of the ergot alkaloids in their biosynthesis. Prior to or during the incorporation of tryptophan into the alkaloid molecule, tryptophan is decarboxylated; for this, the mechanism has not been elucidated at present.

Such incorporation of tryptophan into the ergot alkaloid molecule has been shown to occur in cultures of Claviceps strains which produce the peptide-type (lysergic acid type) ergot alkaloids (7, 60), and in cultures of Claviceps strains which produce predominantly the clavine-type alkaloids.

(C) Feeding methionine-$C^{14}H_3$ to alkaloid-producing Claviceps cultures (8) resulted in incorporation of the C^{14} isotope into the N-methyl group of the alkaloid.

(D) Feeding mevalonic acid-2-C^{14} to cultures of Claviceps (which pro-duced the clavine-type alkaloids) resulted in incorporation of the mevalonic acid into the alkaloids (9, 10, 28, 62). By degradation of the isotope-labelled alkaloids formed in such experiments, it has been shown (9, 10) that the C^{14} isotope from mevalonic acid-2-C^{14} was incorporated into carbon No. 17 of the alkaloid molecule.

In such experiments it was shown (9) that the C^{14} isotope from meva-lonic acid-1-C^{14} was not incorporated into the alkaloid. Also, unlabelled isopentenylpyrophosphate and dimethylallylpyrophosphate were observed (9) to have inhibitory effect on the incorporation of mevalonic acid-2-C^{14} into the

alkaloids. Plieninger et al. (45) fed deuterium-labelled isopentenylpyro-

phosphate to cultures of clavine-alkaloid-producing <u>Claviceps</u> and found in-

corporation of the isotope into the alkaloids formed.

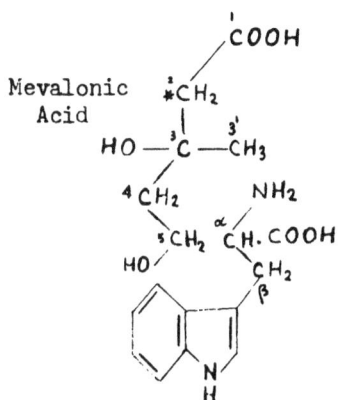

Mevalonic Acid

Tryptophan

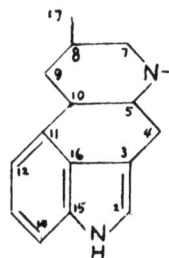

Ergoline Skeleton
of the ergot alkaloids

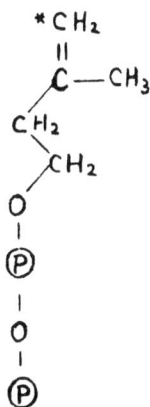

Isopentenyl
pyrophosphate

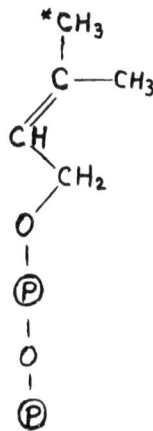

Dimethylallyl
pyrophosphate

These experimental results indicate that mevalonic acid, before becoming incorporated into the ergot alkaloid molecule in the process of biosynthesis, passes through a number of metabolic reaction steps which are
identical with or similar to the several initial reaction steps involving
mevalonic acid in the biosynthesis of sterols (e.g. cholesterol) and terpenoid compounds (e.g. squalene). That is, in its role as precursor for the
appropriate portion of the ergot alkaloid molecule (No. 7, 8, 9, 10 and 17
carbon atoms of the alkaloid molecule), mevalonic acid is probably success-
ively converted to its phosphate, undergoes decarboxylation, and is converted to isopentenylpyrophosphate and/or dimethylallylpyrophosphate before it
is incorporated into the ergot alkaloid molecule.

Thus, the three different component parts of the ergoline skeleton
of the ergot alkaloids are likely to be derived from the three metabolic
precursor sources: (1) The indole nucleus and carbon atoms No. 4 and 5 and
the tertiary nitrogen at position 6 are derived from decarboxylated tryptophan. (2) The N-methyl group may be derived from such metabolic methyl-
donor as methionine. (3) Carbon atoms No. 7, 8, 9, 10 and 17 are derived
from decarboxylated mevalonic acid or more likely from a suitable derivative
of mevalonic acid or some other metabolically suitable isoprenoid compound.
Carbon atoms No. 17, 8, and 7 in the alkaloid molecule correspond to carbons
No. 2, 3, and 3', respectively, of the mevalonic acid molecule.

The mode or mechanism by which these precursor compounds or their
derivatives are joined to form Ring C and Ring D of the ergoline part of
these alkaloids, and whatever intermediate reaction steps that may be necessary in the process of the biosynthesis of the alkaloid molecule, remain un-

solved at present. Certain derivatives of tryptophan have been suggested as possible intermediate compounds as part of the mechanism for the closure of rings C and D (21, 46, 63). Experimental evidence available at present is insufficient for a final conclusion regarding these condensation reactions in the biosynthesis of the ergot alkaloids.

8.2 Alkaloids of Rauwolfia

(a) Plant Sources

The roots and leaves of various Rauwolfia species have been used for a wide variety of medical purposes in India for centuries. Chemical and pharmacological studies on Rauwolfia plant materials and their alkaloids have been carried out only in the past three or four decades. Since the introduction of Rauwolfia serpentina Benth (Fam. Apocynaceae) and its alkaloid reserpine into Western medicine some twenty years ago, the increase of their medicinal use as hypotensive and sedative agents has stimulated considerable interest in them and has produced a voluminous chemical and pharmacological literature during the past two decades.

The root is the principal organ in which the alkaloids occur in the plant of the various Rauwolfia species. Over twenty different species, collected from tropical regions of Asia, Africa, and Central and South America, have been investigated chemically by many different research groups. Some sixty different alkaloids have been isolated from these plants (13, 38). Rauwolfia serpentina is at present the only species of which the root material constitutes or enters into certain medicinal preparations as such or in the form of an alkaloids-containing extract, and it is the only species at present recognized in the N.F. XII and the B.P. 1963. However, many species of *Rauwolfia* other than *R. serpentina* are also of considerable interest from the chemical and medicinal points of view as they serve as important raw material sources for the isolation and commercial production of the three individual alkaloids which are by themselves in the pure state used as medicinal agents. The three medicinally most important, naturally occurring alkaloids occurring in various *Rauwolfia* species are: reserpine, deserpidine, and rescinnamine. A few of the more important plant species which often serve as the raw material sources for the isolation or production of these three alkaloids are listed in Table 8.

The following discussions on the chemical aspects of the *Rauwolfia* alkaloids will be mainly concerned with these three alkaloids but will include some of the other alkaloids occurring in *Rauwolfia* to be discussed in relation to these three medicinally useful ones.

The data given in Table 8 have been adapted from those given by Woodson (64) and by Lucas (38). There are, of course, many more alkaloids which have been isolated from these species; in some cases, a score or more individual alkaloids have been isolated from one particular single species. There are also other *Rauwolfia* species in which these and other related alkaloids occur. The six principal alkaloids listed in Table 8 are those with which our discussion of the chemical aspects will be chiefly concerned.

The literature contains numerous reports of investigations carried out on *Rauwolfia canescens* L., *R. heterophylla* Willd. ex Roem. et Schult, *R. hirsuta* Jacq. All these three species are now considered to be identical with *Rauwolfia tetraphylla* L. (38, 64), and therefore in Table 8 only one of these four names is listed.

Table 8. Some Rauwolfia Species and
Their Principal Alkaloids

Species name	Principal geographical distribution	Principal Alkaloids
R. serpentina Benth. ex Kurz	Indian subcontinent, Ceylon, Java, Burma, Thailand	Reserpine, Ajmaline, Serpentine, Rescinnamine
R. vomitoria Afzel	Guinea, Uganda to Angola, Nyasaland	Reserpine, Ajmaline, Rescinnamine
R. tetraphylla L. (see text for other names)	Mexico to Venezuela, Ecuador, Antilles	Deserpidine, Reserpine, Aricine, Serpentine, Ajmaline.
R. micrantha Hook f.	S.W. India	Serpentine, Reserpine
R. perakensis King et Gamble	Burma, Malaya, Indochina	Reserpine
R. sellowii Muell.-Arg.	Brazil	Reserpine, Aricine, Ajmaline, Serpentine
R. densiflora Benth. ex Hook f.	India, Ceylon, Burma, Malaya, Java	Reserpine, Ajmaline
R. schueli Speg.	Bolivia, Argentina	Reserpine, Aricine
R. nitida Jacq.	Antilles	Reserpine, Rescinnamine

A content of 0.15 - 0.2% alkaloids of the reserpine group is specified for the Powdered Rauwolfia serpentina N.F. XII and B.P.C. 1963. The

reserpine content of some __Rauwolfia__ species other than R. serpentina is considerably higher than 0.2%, especially in R. vomitoria roots. Total alkaloid content of R. serpentina may be as high as 0.8% or higher. R. sellowii has been reported to contain as much as 1.2% ajmaline (13). Ajmaline is the most abundant alkaloid in R. serpentina.

(b) Grouping of Rauwolfia Alkaloids

All the Rauwolfia alkaloids are indole bases. Schlittler (64) has divided the Rauwolfia alkaloids into three groups:

I.   Yellow-coloured, strong anhydronium bases, pK_a 10.4 - 11. e.g., serpentine, serpentinine, alstonine.

II. Colourless tertiary indoline alkaloids of intermediate basicity, pK_a 8.15 - 8.3. e.g., ajmaline, isoajmaline, rauwolfinine, semperflorine, tetraphyllicine.

III. Colourless weakly basic tertiary indole alkaloids, pK_a 6 - 7.5. This group may be subdivided into two sub-groups:

(a) those with heterocyclic ring E, e.g., aricine, reserpinine, ajmalicine (δ-yohimbine), raubasine, tetrahydroserpentine.

(b) those with carbocyclic ring E., e.g., reserpine, deserpidine, rescinnamine, yohimbine, serpine, sarpagine.

In addition to the pK_a values, each of these groups of Rauwolfia alkaloids also gives ultraviolet spectrum which is rather indicative of the group (64).

(c) Structures and Properties of the Principal Alkaloids

As may be seen from the structure formulae, the three most important medicinal alkaloids of Rauwolfia, reserpine, deserpidine, and rescinnamine,

- 116 -

are diesters.

In <u>reserpine</u>, the methyl ester of the pentacyclic reserpic acid is linked, through its hydroxyl group at position 18, by an ester linkage, to trimethoxybenzoic acid.

<u>Deserpidine</u>, which has also been called canescine, recanescine, raunormine, and 11-desmethoxyreserpine, has the same structure as reserpine except that the methoxyl group at position No. 11 in reserpine is replaced by a hydrogen atom in deserpidine. That is, in deserpidine, the methyl ester of deserpidic acid (the pentacyclic part) is linked through its hydroxyl group at position 18 by an ester linkage to trimethoxybenzoic acid.

In <u>rescinnamine</u>, the methyl reserpate is linked by ester linkage to trimethoxycinnamic acid, instead of trimethoxybenzoic acid.

Therefore, in all these three alkaloids (which are all laevorotatory), on alkaline hydrolysis, the two ester linkages in each case are split. This gives trimethoxybenzoic acid, reserpic acid and methyl alcohol in the case of reserpine; or trimethoxybenzoic acid, deserpidic acid and methyl alcohol for deserpidine; or trimethoxycinnamic acid, reserpic acid and methyl alcohol in the case of rescinnamine. Such hydrolysis may be effected by using 0.75 N or 0.9 N sodium hydroxide in methanol: water 3:1 (36, 37). The products of hydrolysis may be separated, for example, in the case of rescinnamine, by acidifying the solution with hydrochloric acid to pH 2 and then extracting with chloroform. The chloroform extract may then be concentrated in vacuo and the trimethoxycinnamic acid crystallizes out from it. From the aqueous portion, standing in refrigerator overnight, the reserpic acid crystallizes out as its hydrochloride (36, 37).

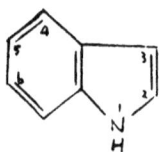

Indole

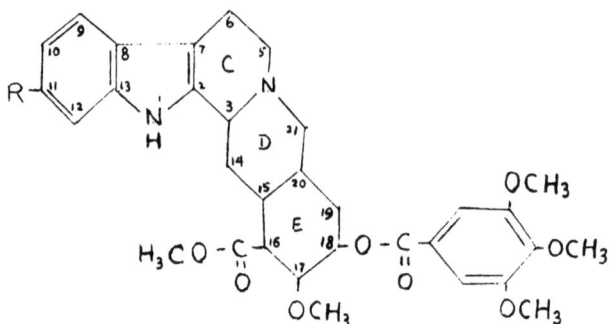

Reserpine   R = -OCH$_3$
Deserpidine   R = H

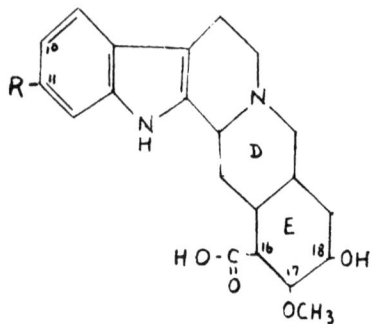

Reserpic Acid R = -OCH$_3$
Deserpidic Acid R = H

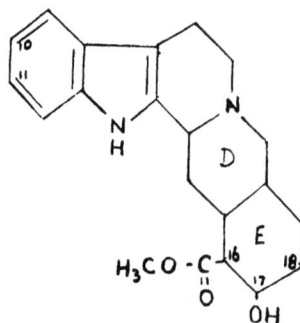

Yohimbine

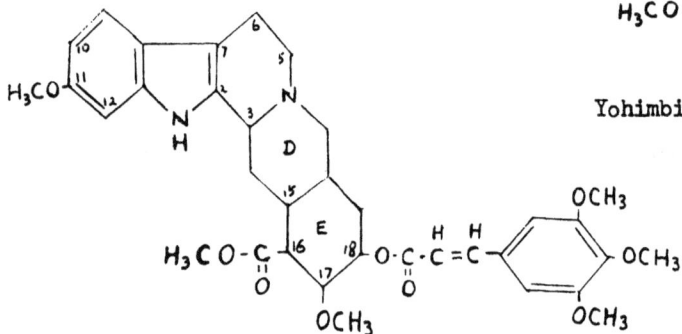

Rescinnamine

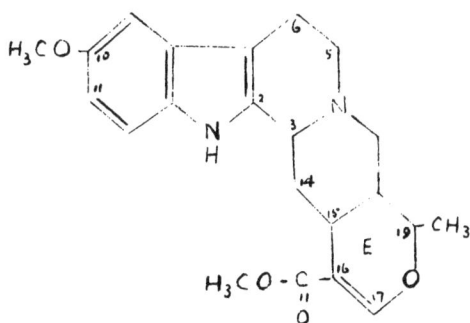

Aricine

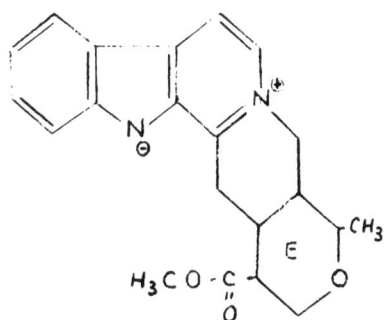

Serpentine

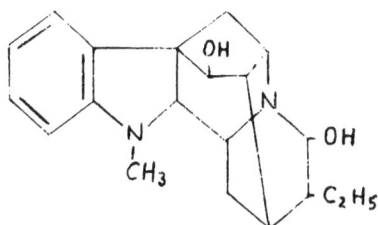

Ajmaline

When treated with lithium aluminum hydride, reserpine yields 3,4,5-trimethoxybenzyl alcohol and reserpic alcohol (reserpine-diol). Reserpic alcohol is reserpic acid with its COOH group replaced by CH_2OH (42).

Reserpic acid may be esterified with methanol and hydrochloric acid to yield methyl reserpate. Acylation of methyl reserpate with trimethoxy-benzoyl chloride (trimethylgalloyl chloride) in pyridine forms reserpine.

Reserpine, deserpidine, and rescinnamine belonging to the same group of Rauwolfia alkaloids in Schlittler's grouping (the weak tertiary indole bases) have rather similar ultraviolet spectra. That for reserpine has maxima at 215, 267, and 295 mµ. with shouder at 225;   the U.V. spectrum for

deserpidine has maxima at 216, 271 and 289 mµ.   These are indicative of the

2,3-disubstituted indole and the 3,4,5-trimethoxybenzoate chromophores in

their molecules.   Rescinnamine has its U.V. spectrum with maxima at 229 and

302 mµ. with a minimum at 258 mµ.   The peak at 302 mu. is interpreted as

arising from the combined effects of the two chromophores, (a) 2,3-disubsti-

tuted-6-methoxyindole, and (b) 3,4,5-trimethoxycinnamic acid (13, 36, 37).

(d) Isolation of Individual Alkaloids

A total alkaloid extract may be obtained from powdered root material

of an appropriate Rauwolfia species by extraction with a suitable solvent.

Ethanol and ethylene chloride are frequently used.   The total alkaloid ex-

tract is then usually chromatographed on column (acid washed alumina is

widely used).   Benzene is often used as the elution solvent for reserpine,

deserpidine, and rescinnamine.   In a chromatographic elution sequence going

from benzene to a degree of solvent polarity as represented by 10% methanol

in chloroform, the sequence of the principal Rauwolfia alkaloids eluted is

represented as follows: (13).

10% methanol	Sarpagine
in chloroform	Serpentine
	Serpentinine
↑	Ajmaline
	Yohimbine and isomers
	Rescinnamine
	Reserpine
	Tetrahydroserpentine
benzene	Reserpinine

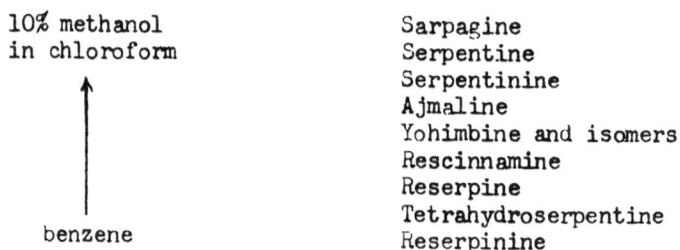

Isolation of Reserpine and Rescinnamine

In 1947, in connection with the estimation of total alkaloids by

conventional solvent extraction of Rauwolfia serpentina, Dutt et al. (17)

made an extract of the powdered root material with alcohol. The solvent was then distilled off from the alcoholic extract, leaving a dry residue. This residue was then treated with several changes of water, after which there was left some insoluble matter (i.e., water-insoluble material from the alcohol-soluble extract) and this was designated as "oleoresin". A number of other investigators, a few years later, found reserpine in this "oleoresin" fraction.

During that period, there was also commercially available a fat-soluble alkaloidal fraction extracted from R. serpentina root known as alseroxylon (which is also in use at present in a number of medicinal preparations of Rauwolfia). A number of investigators also isolated reserpine in the pure state from this alseroxylon extract.

Reserpine was isolated in the pure state by Müller, Schlittler, and Bein in 1952 (41) from the "oleoresin fraction". Independently, Klohs et al. (35, 36, 37) isolated reserpine from the so-called "oleoresin fraction", and obtained crystalline reserpine and crystalline rescinnamine from the alseroxylon preparation. Their procedure (36, 37) for isolating reserpine and rescinnamine from the alseroxylon fraction may be summarized in outline as shown in chart-form in Figure 6.

Isolation of Deserpidine

Deserpidine was isolated almost simultaneously in 1955 by four different groups of investigators (hence several different names were given to it) from R. canescens (R. tetraphylla) (13, 64). This species is particularly rich in deserpidine although it also occurs in a number of other species in smaller quantities. In the procedure used by Neuss et al. (42), after crystallization of reserpine from the benzene extract, the mother liquor is reduced, in vacuo, to a residue. A 7.4 gm. sample of such a mother-liquor residue dissolved in 50 ml. of benzene is chromatographed on 300 gm. of acid washed alumina column, using benzene as eluent. The first fourteen 100-ml. fractions are collected and evaporated under reduced pressure to a residue of 1.4 gm. of deserpidine (recanescine) which is then crystallized from ethyl acetate (the crystals containing 1 mole of solvent).

(e) Pharmacological Action and Uses

Rauwolfia and its alkaloids reserpine, deserpidine and rescinnamine have multiple sites of action. The mechanism of pharmacological action for these Rauwolfia alkaloids is not entirely elucidated.

They produce depressor effect (hypotensive action) probably due to vasodilatation. They are widely used either alone or in combination with other more powerful depressor agents as antihypertensive drugs.

Rauwolfia and its alkaloids (the three just mentioned) also possess tranquilizing properties and are used as tranquilizers in the management of various states of anxiety, apprehension and psychomotor agitation.

Reserpine has been shown to have the property of releasing serotonin (5-hydroxytryptamine) from its bound form in tissue depots particularly in the brain. The serotonin thus released is then subject to destruction by monoamine oxidase and other enzymes in the body tissues. The exact relationship between this action and its hypotensive and/or sedative effects is still not fully understood.

Alseroxylon, which is a fat-soluble alkaloidal fraction extracted from Rauwolfia serpentina root and contains reserpine and other non-adreno-

Alseroxylon (100 gm.)
dissolved in 170 ml. MeOH
↓
Stir with 3.5 L. benzene for 1 hr.
↓
Filter

Solid
↓
Repeat stirring with
benzene and filter ⟶ Filtrate (benzene ext.)
↓
Concentrate in vacuo to 4.8 L.

Yellow ppt.   Filtrate
(removed)   ↓
Reduce to dryness in vacuo
(alkaloids, tannins, resins)
↓
Dissolve in 150 ml. of 2.5%
acetic acid in MeOH
↓
Add ammonia to pH 8.5
↓
Standing

Mother   Reserpine
liquor   crystallized
↓   out (5.3 gm.)

Evaporate in vacuo to
dryness (5 gm. residue)
↓
Dissolve in minimum amt.
of benzene
↓
Chromatographed on 100 gm.
acid-washed alumina
↓
Fractions eluted with sequence of solvents from chloroform through chloroform-
1% MeOH
↓
Rescinnamine (0.7 gm.)
(crystallized from benzene)

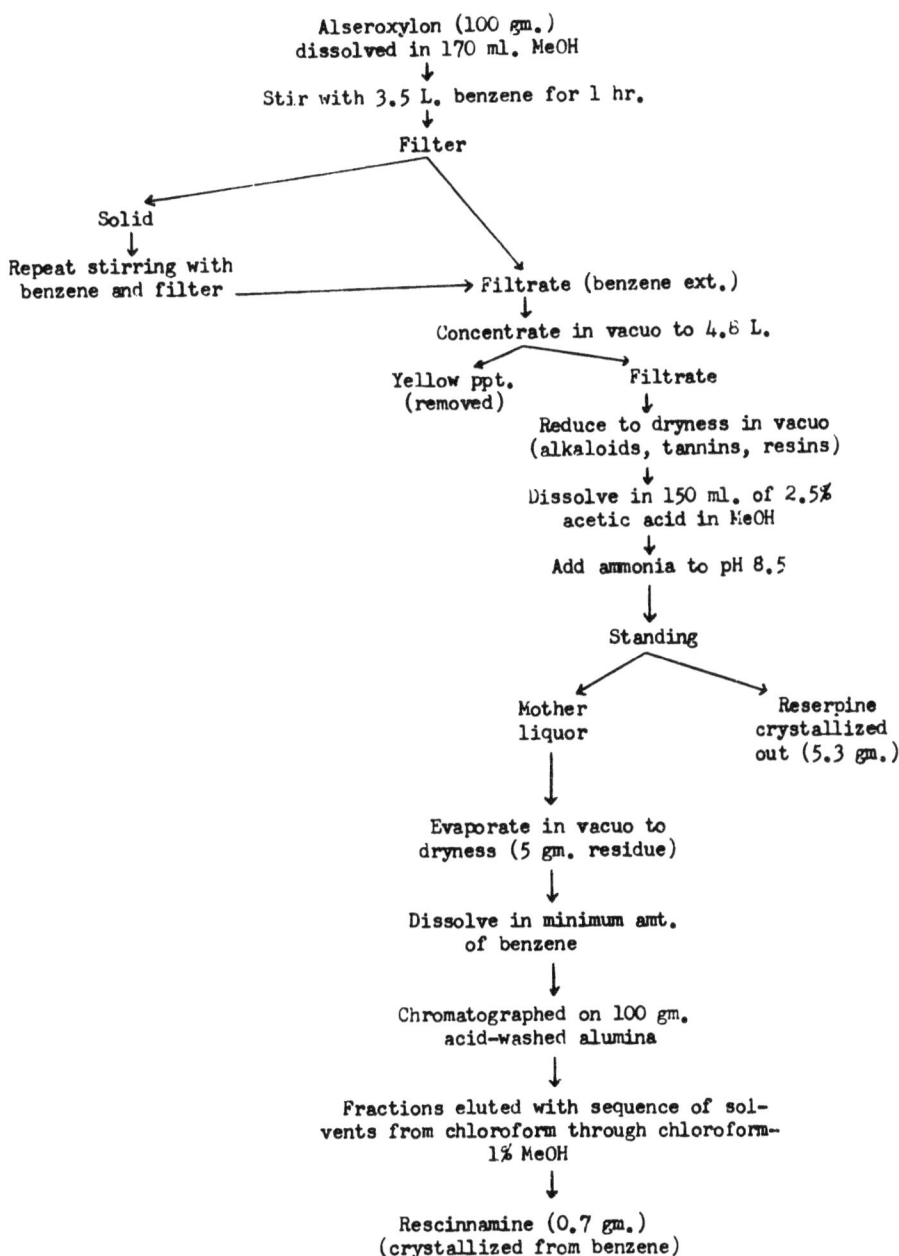

Figure 6. Procedure Outline for Isolation of Reserpine and Rescinnamine

lytic alkaloids, exerts a pharmacological effect approximately equivalent to that produced by one-fifth of its weight of pure reserpine.

(f) Medicinal Preparations

     Rauwolfia Serpentina N.F. XII

     Rauwolfia B.P.C. 1963

     Reserpine U.S.P. XVII, B.P.C. 1963

     Rescinnamine N.F. XII

     Syrosingopine N.F. XII (synthetic derivative of reserpine)

Examples of Pharmaceutical Specialties

     (a) Those containing R. serpentina root material:
          Raudixin[R]; Mio-Pressin[R] (with Veratrum alkaloids).

     (b) Those containing alkaloidal extracts of R. serpentina root:
          Rautensin[R]; Rauwiloid[R]; Rauvera[R] (with Veratrum extract).

     (c) Those containing purified reserpine:
          Eskaserp[R]; Rau-Sed[R]; Reserpoid[R]; Roxinoid[R];
          Serfin[R]; Serpasil[R]; Serpiloid[R]; Sandril[R].

     (d) Harmonyl[R] (deserpidine).

     (e) Moderil[R] (rescinnamine).

     (f) Singoserp[R] (Syrosingopine) (reserpine derivative).

8.3 Alkaloids of Nux Vomica

(a) Plant Sources

     A number of alkaloids which are dihydroindole derivatives occur in several species of Strychnos. The best known among these alkaloids are strychnine and brucine which occur in the endosperm of the seeds of Strychnos nux-vomica L., Strychnos ignatii Berg. (Fam. Loganiaceae) and several other Strychnos species in varying quantities from 0.2 to 5%. Nux Vomica

(the seeds of S. nux-vomica) of the B.P.C. 1963 is specified to contain not

less than 1.2% of strychnine. The bark of some of these species contains a

considerable quantity of both of these alkaloids, especially brucine. A

brucine content of 7.78% in the bark of S. nux-vomica has been reported

(15b). The leaf of this species has been reported to contain 1.8% of total

alkaloids and 0.7% of strychnine (48).

(b) Structures and Properties

Although strychnine was isolated in 1817, its complex structure for-

mula as shown below with a high degree of certainty based on experimental

evidence has been worked out only recently, during the period 1946 - 1949,

largely by Woodward and his group at Harvard and Prelog and his group at

Zurich and Robinson and his group at Oxford (32).

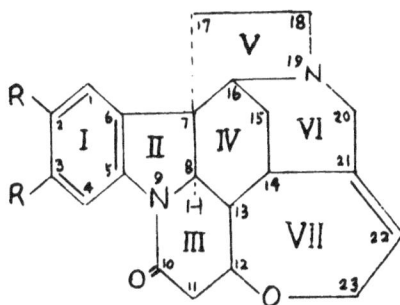

Strychnine   R = H
Brucine   R = -OCH3

Properties

Strychnine and brucine behave as monoacidic bases of intermediate

basicity, with a pK_a of 7.37 and 7.45 respectively. They form various crystalline salts. Their solubility data are listed in Table 9.

Table 9. Solubility Data of Strychnine and Brucine

| | No. of ml. of solvent to dissolve 1 gm. at 25 | | | | | |
	Water	Ethanol	Absolute ethanol	Chloro- form	Benzene	Amyl alcohol
Strychnine	6400	160 (90% EtOH)	350	6	160	180
Strychnine Sulphate $B_2.H_2SO_4.5H_2O$	31	65		325		
Strychnine Nitrate $B.NHO_3$	42	120		156		
Strychnine Hydriodide	spar- ingly					
Brucine	320	v. sol.		v. sol.		
Brucine Sulphate $B_2.H_2SO_4.7H_2O$	solu- ble					

The ferrocyanide, chromate, and hydriodide of strychnine are less soluble in water than the corresponding salts of brucine.

The nature of the component parts of the molecular structure of strychnine (and of brucine) gives rise to certain products of oxidation, hydrolysis and other degradative processes. In fact, before the present structure formula for strychnine was known, it was the identity of the products of degradation by various methods which provided the clues leading to the final formulation of the structure formula. A few of the basic and less complex ones of these degradative processes may be briefly summarized as follows:

The two methoxyl groups (positions 2 and 3) of brucine may be split off by heating the alkaloid in hydrochloric acid in a sealed tube.

Catalytic hydrogenation of strychnine (or of brucine) reduces the $C_{21}-C_{22}$ double bond, producing the dihydro alkaloid.

It may be noted that rings I and II form the dihydroindole nucleus of strychnine. Positions 8 and 7 in the numbering system shown above for the strychnine molecule correspond to what would be called positions 2 and 3 (or alpha and beta positions) in the molecule of indole. Therefore, the strychnine molecule contains an α,β –disubstituted–dihydroindole.

On alkaline pyrolysis, strychnine yields indole as one of the products. Treatment of strychnine with alcoholic potassium hydroxide yields, among other products, tryptamine (derived from positions 1 – 9 and 17, 18, and 19).

Oxidation of strychnine by hot 20% nitric acid yields dinitrostrycholcarboxylic acid, 3,5–dinitro–benzoic acid, and picric acid. This is because of the presence of the α,β –disubstituted–N–acyl–dihydroindole component in the strychnine molecule (32). Alkaline permanganate oxidation of strychnine yields oxalylanthranilic acid (among other products) and this is because of the nature and relative positions of rings II and III in the strychnine molecule.

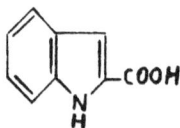

Strychol

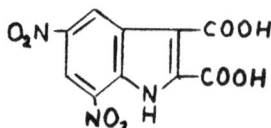

Dinitro-strychol-
carboxylic acid

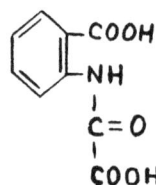

Oxalylanthranilic
acid

(c) Colour Tests

A very small quantity of strychnine in a drop of 80% sulphuric acid, when stirred with a crystal of potassium dichromate gives reddish violet to purple coloration. Manganese dioxide or potassium permanganate may also be used instead of the dichromate for this test. This test is said to be extremely sensitive and is given by very few other alkaloids (except gelsemine, yohimbine and protocurarine) (15b). Strychnine salts will also give this test, except strychnine nitrate.

Strychnine or its salt when treated with sulphuric acid and ammonium vanadate (or with a saturated solution of ammonium vanadate) gives a violet to blue coloration. Under certain standardized conditions, this coloration may be produced by less than one μg. of strychnine (14).

Strychnine when treated with a trace of nitric acid gives a yellow colour, while brucine gives an intense orange-red colour with a trace of nitric acid.

(d) Isolation of Strychnine and Brucine

The plant seeds after being ground to a powder are mixed with slaked

lime and water to make a paste, which is then dried at 100°, and then extracted with hot chloroform (by continuous extraction procedure). The
chloroform extract is shaken with dilute sulphuric acid (into which the alkaloids pass in the form of their salts). The aqueous acid extract is made
alkaline with ammonia, thereby the water-soluble alkaloid salts are converted back to their free base forms which are only very sparingly soluble
in water and which thus come down as precipitates.

The alkaloidal precipitates are extracted with 25% alcohol, which
dissolves brucine and leaves strychnine as insoluble residue. After separation by filtering, the residue (strychnine) is purified by repeated crystallization of the strychnine from alcohol. The solution, containing brucine,
is concentrated and then neutralized with oxalic acid when brucine and impurities are precipitated as oxalates. The precipitates are washed with
cold alcohol and then dissolved in hot water, and decolourized by charcoal.
The hot aqueous solution is mixed with magnesia and evaporated to dryness on
a water bath. Brucine is then extracted from the residue by acetone, the
acetone extract evaporated to dryness, and brucine in the residue is purified by repeated crystallization from dilute alcohol (15b).

(e) Quantitative Estimation of Strychnine

(i) Extraction and Titration: - After appropriate extraction of strychnine from the plant material (usually by chloroform and dilute sulphuric
acid similar to the isolation procedure described above), the organic solvent extract containing the alkaloid base is then dissolved in a known quantity of a standardized acid (usually 0.02 N HCl) in slight excess of what is
required to combine with the quantity of alkaloid present, and the excess

acid is then titrated with a standardized alkali hydroxide solution and the amount of alkaloid calculated from the acid equivalents. As pointed out above, strychnine behaves as a monoacidic base in its formation of salt with acid. The selection of an indicator for such titration is governed by the pK_a value of the alkaloid. This is essentially the basis of the assay procedure for strychnine in Nux Vomica in B.P. 1963.

(ii) <u>Adsorption Chromatography and Titration</u>: - A chromatographic method of separation of the alkaloids has been worked out by Fischer and Buchegger (15b). In this procedure, a 2-gm. sample of powdered seed material (made alkaline) is extracted with chloroform and the chloroform extract evaporated to dryness. The residue is dissolved in trichloro-ethylene and chromatographed on alumina (10 gm.) column. Strychnine is then eluted with 70 ml. of carbon tetrachloride containing 9% acetone, while brucine is eluted with 25 ml. of alcohol. In each case, the solvent is then removed by distillation and the alkaloidal residue titrated with a standard acid.

(iii) <u>Separation by Ion Exchange Resin</u>: - In the procedure of Jindra and Pohorski (33), a 0.3 gm. sample of the plant material (Nux Vomica) is extracted with a mixture of 5 gm. of chloroform, 15 gm. of ether and 1 ml. of 10% ammonia by shaking. The chloroform-ether extract is then shaken for thirty seconds with 5 ml. of water and the chloroform-ether layer is then evaporated to almost dryness, 1 ml. of alcohol added, and evaporation is carried to dryness. To the residue is then added three drops of 2% sulphuric acid, followed by 10 ml. of 90% alcohol and the solution is passed through ion exchange resin (Amberlite IR-4B), to remove impurities, and the eluate is titrated with 0.01 N hydrochloric acid, using the antimony electrode for electrometric titration, and the quantity of strychnine calculated.

(iv) Strychnine may also be quantitatively estimated by ultraviolet spectrometric method (19). This makes use of the absorption band at 254 mμ. shown by strychnine in absolute alcohol (brucine in absolute alcohol shows maxima at 264 and 301 mμ. in its U.V. spectrum): The alkaloid extract (containing both strychnine and brucine) is treated with 0.5 gm. of potassium persulphate in the presence of 3% sulphuric acid for one hour, thereby brucine is destroyed while strychnine is not. The solution is then made alkaline and the strychnine extracted with chloroform. The residue (containing strychnine) left after removal of solvent from the chloroform extract is dissolved in absolute alcohol and the extinction (E) at 254 mμ. of the solution measured in a suitable spectrophotometer. This E value divided by $E_{1\%}^{1cm}$ at 254 mμ. (which is equal to 390) gives the concentration, as per cent, of strychnine in the alcoholic solution being measured. This procedure may be carried out on pharmaceutical preparations such as liquid extract and tincture of Nux Vomica in 2 ml. samples. Impurities which may absorb at 254 mμ., if not properly removed by the extraction would of course interfere with accurate quantitative estimation of the strychnine. With proper removal of such impurities, it is claimed that the average error of this procedure carried out on known samples was found to be very small (0.07%).

(f) Pharmacological Action and Uses

Strychnine exerts convulsant action mainly by depressing the transmission of inhibitory fibres which innervate the motor horn cells in the spinal cord. Vasomotor centres in the medulla oblongata may also be stimulated by strychnine. Today, most pharmacologists consider its former use as a tonic as rather irrational. It is now seldom used as a therapeutic agent, but is used rather as a tool in conjunction with various pharmacological investigations related to centres in the spinal cord.

(g) Medicinal Preparations

Nux Vomica B.P.C. 1963

Nux Vomica Tincture B.P.

Strychnine Hydrochloride B.P.C.

8.4 Physostigmine

The seeds of Physostigma venenosum Balfour (Fam. Leguminosae), known as Calabar Bean or Ordeal Bean, contain several alkaloids which are indole derivatives. Physostigmine, which is also known as eserine, is the principal one among them. It occurs in Calabar bean to the extent of 0.15 - 0.3%.

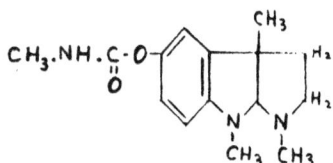

Physostigmine
(Eserine)

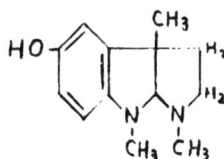

Eseroline

(a) Properties and Reactions

Physostigmine is a monoacidic tertiary base, soluble in alcohol, chloroform, and benzene. It is readily oxidized by oxygen in the presence of potassium hydroxide solution to a red compound, rubreserine. Hydrolysis of physostigmine by alkali gives methylamine, carbon dioxide, and eseroline (the methylamine and carbon dioxide being derived from the carbamate of the urethane side chain of physostigmine). Oxidation of physostigmine with potassium permanganate yields methyl isocyanate. Heating the alkaloid with sodium ethylate in the absence of air gives methyl urethane ($CH_3.NH.CO_2.$ C_2H_5).

The degradation product eseroline can be reconverted to physostigmine by methyl isocyanate in the presence of a trace of sodium and atmospheric oxidation of eseroline gives rise to the red compound rubreserine (39b).

Physostigmine gives colour reactions with a number of reagents. For example, with a solution of phosphomolybdic acid and ammonium meta vanadate in sulphuric acid, physostigmine produces an emerald green colour. It gives a blue colour reaction with potassium ferricyanide and ferric chloride.

When physostigmine is heated in potassium hydroxide solution, a deep yellow colour is produced. It is stated that this colour reaction can detect as little as 10 μg. of the alkaloid. Under controlled conditions, the intensity of this colour may be measured spectrophotometrically at 470 mμ. and serves as a quantitative determination (50b). Physostigmine is stated to be precipitated quantitatively by sodium tetraphenyl borate (50b).

(b) Extraction of Physostigmine

The powdered seed material may be extracted by continuous percolation with hot alcohol, and the solvent is then removed by distillation. Addition of water to the residue and separation of the floating fat layer are followed by alkalinization of the aqueous liquid with sodium carbonate,

and repeated extraction with ether. The ether extract is then concentrated
to a small volume and washed repeatedly with 5% sulphuric acid until washings just become acidic. To the aqueous acid solution (containing the alkaloid sulphate) is added an excess of a saturated solution of sodium salicylate, thereupon crystalline physostigmine salicylate is precipitated out
(39b).

(c) Pharmacological Action and Uses

Physostigmine produces parasympathomimetic effects by its inhibition
of cholinesterase. It is thus a stimulant of the organs innervated by cholinergic nerve fibres, producing accordingly miosis, increase of tone and peri-
stalsis of the gastrointestinal tract, and stimulation of secretions such as
saliva, pancreatic juice and perspiration. Physostigmine counteracts the
paralyzing effects of nicotine and curare alkaloids on ganglia of the autonomic nervous system.

It is used in glaucoma chiefly for its action in reducing the intra-
ocular pressure. It has also been used in myasthenia gravis, but for this
purpose it is now largely replaced by neostigmine, which is one of the many
synthetic compounds made and tested for such physiological activity as a result of the discovery of the importance of the urethane side chain in physostigmine to its pharmacological action. The degradation product eseroline,
which lacks the urethane side chain of physostigmine, loses physostigmine's
physiological activity.

Medicinal Preparations

       Physostigmine B.P.C. 1963

       Physostigmine Salicylate U.S.P. XVII, B.P.C. 1963

2.5 Vinca Alkaloids

(a) Plant Sources

Since the middle 1950's, within a decade, some sixty alkaloids have been isolated from the tropical ornamental periwinkle plants, especially Catharanthus roseus (L.) G. Don, Vinca major L., Vinca minor L., and several other related species (Fam. Apocynaeae) (20, 57). The species Catharanthus roseus (L.) G. Don has also been referred to in the literature as Vinca rosea L. and as Lochnera rosea (L.) Reichenbach. Plant taxonomists appear to be generally in agreement that Catharanthus roseus and other species which have now been placed in the genus Catharanthus (or Lochnera) are generically distinct from other Vinca species and should therefore be placed in a separate genus with a name different from the genus name Vinca. The species Vinca major L., Vinca minor L., and certain other related species properly belong to the genus Vinca (and not Catharanthus or Lochnera), where-as Catharanthus roseus (Lochnera rosea) and other Catharanthus (Lochnera) species do not belong to the genus Vinca. Botanists are not in agreement as to whether Catharanthus should be given priority over Lochnera, or vice versa, as the name of the genus according to the rules of botanical nomenclature (18). The alkaloids isolated from the Catharanthus (Lochnera) species and from the Vinca species are collectively referred to as the Vinca alkaloids.

(b) Structures and Properties

Among these Vinca alkaloids, vinblastine (formerly known as vinca-leukoblastine), vincristine (formerly known as leurocristine), vinleurosine (formerly leurosine), and vinrosidine (formerly leurosidine) are the best known ones with respect to laboratory and clinical tests of them as oncolytic agents. The chemistry and biological activities of these and other Vinca alkaloids have been intensely studied in the past few years.

Vinblastine (pK$_a'$ 5.4, 7.4) and vincristine (pK$_a'$ 5.0, 7.4) are unsymmetrical dimeric alkaloids containing indole and indoline (dihydroindole) moieties in the molecule (43, 57). There are also a number of monomeric alkaloids (24) isolated from Catharanthus roseus and other related species. Some of these monomeric Vinca alkaloids contain an indole moiety (e.g., catharanthine, lochnerine, etc.), while some other monomeric Vinca alkaloids contain an indoline moiety (e.g., vindoline, vindolinine, etc.) (24, 25). These Vinca alkaloids (dimeric and monomeric types) seem to occur in all parts of the plant. As may be seen from the extraction and isolation procedure schemes described in outline below, the quantity of vinblastine occurring in the plant (Catharanthus roseus) is only of the order of 0.003% of dry whole-plant material (58) or of the order of 0.005% of the dry leaf material (44). Many of the minor alkaloids occurring in these plants were isolated in yields of 0.0001% or less. Catharanthine, vindoline, and vindolinine are among the relatively few major alkaloids of the leaf of Catharanthus roseus.

Vinblastine was isolated in 1958 by Noble, Beer and Cutts (44) as the sulphate in crystalline form. Vinblastine sulphate is soluble in water and in methanol, and sparingly soluble in ethanol. It crystallizes from ethanol as small colourless needles. The free base may be prepared from the sulphate by basifying the solution of the sulphate and extracting with methylene chloride. Evaporation of the methylene chloride extract so obtained leaves the free base form of vinblastine as an amorphous white powder (44).

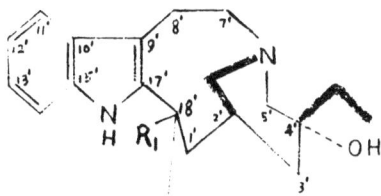

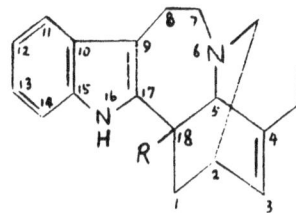

Catharanthine
R = COOCH$_3$

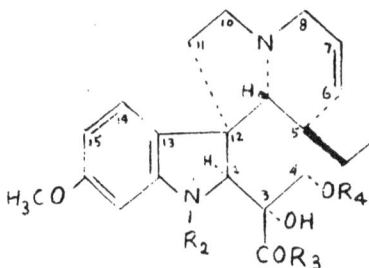

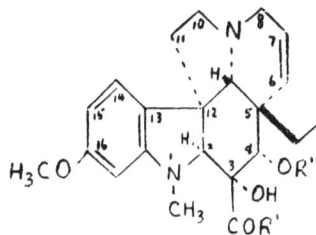

	Vincristine	Vinblastine	Vindoline
			R' = OCH$_3$
			R" = COCH$_3$
R$_1$	COOCH$_3$	COOCH$_3$	
R$_2$	CHO	CH$_3$	
R$_3$	OCH$_3$	OCH$_3$	
R$_4$	COCH$_3$	COCH$_3$	

(c) Extraction and Isolation of the Oncolytic Alkaloids

(i) Isolation of Vinblastine (Beer's Method): – In the method of

Noble, Beer and Cutts (44), the dried leaf material is extracted with hot

ethanol:water:acetic acid (9:1:1). After removal of solvent, the residue is

extracted with hot 2% hydrochloric acid, and the acid extract is then ad-

justed to pH 4 when certain non-alkaloidal precipitates separate out and are removed by centrifugation. The aqueous acidic solution (the supernatant) is then adjusted to pH 7 and extracted with benzene. The dry residue obtained from evaporation of this pH 7 benzene extract (1.21 gm. derived from 600 gm. of dried leaf material) contains most of the vinblastine together with other alkaloids. Phenolic and tarry materials are then removed from this alkaloidal extract by washing with dilute alkali. A 1.38 gm. sample of the washed extract (the pH 7 benzene extract) is chromatographed on 120 gm. of Woehlm Grade IV-V alumina and elution is carried out in 18 fractions, starting with benzene-methylene chloride (65:35) mixture, and progressively diluted with methylene chloride to pure methylene chloride for the last fraction. Vinblastine is largely recovered from Fraction 9 which yields 27.5 mg.

(ii) <u>Extraction and Separation (Svoboda's Method)</u>: - Svoboda et al. (56, 57, 58) have devised an extraction scheme by which numerous individual Vinca alkaloids have been isolated. The plant material (ground whole plant, <u>Catharanthus roseus</u>) after defatting with Skelly B (essentially n-hexane), is extracted with 2% tartaric acid before extraction with organic solvents under acidic and under alkaline conditions, thus initially separating those alkaloids whose tartrates are soluble in organic solvents and those which are not. Some alkaloids are also removed from the plant material by the Skelly B together with fat. As this extraction scheme and the subsequent chromatography and gradient pH extractions may serve to illustrate recent advances from classical extraction techniques (those discussed in Chapter 2), and the principles are applicable to the separation of certain other groups of alkaloids, these will now be described in some detail.

In the following outline of the Svoboda extraction and separation
schemes, only the details pertaining to those steps and fractions which
yield vinblastine, vincristine, vinleurosine, and vinrosidine are summarized,
and the details for those other fractions yielding a number of Vinca alkaloids other than these four are mostly excluded.

The initial steps of extraction are summarized schematically in
Figure 7. Each of the Fractions A, A_1, B, B_1, etc.) is then chromatographed
on alumina, and eluted with various eluting solvents. Some of these column
fractions (i.e., eluted from the alumina column) are further subjected to
gradient pH extractions to achieve separation of certain ones of the Vinca
alkaloids.

In this scheme, vinblastine, vinleurosine, vincristine, and vinrosidine are in the Fraction A, together with a number of other alkaloids.

Chromatography: - A benzene solution of 10 gm. of the extraction
Fraction A is chromatographed on 400 gm. of Alcoa Alumina Grade F-20 (deac-
tivated by treatment with 12.5 ml. of 10% acetic acid). Eluting solvents
used are benzene, benzene:chloroform, chloroform, and chloroform:methanol,
in 500-ml. fractions. Vinleurosine (0.234 gm.) comes down in fractions 34 -
42 (benzene:chloroform 1:1); vinblastine (0.126 gm.) is obtained as the
sulphate from fractions 43 - 45 (benzene:chloroform 1:1) (58).

The post-vinblastine eluent fractions (3.6 Kg. of residue from evaporation of combined chloroform fractions) are re-chromatographed on deacti-
vated alumina (120 Kg.) in a similar manner. The eluting solvents and fractions collected are: benzene (Fraction 1); benzene:chloroform 3:1 (Fractions 2 - 15); benzene:chloroform 1:1 (Fractions 16 - 29); benzene:chloro-

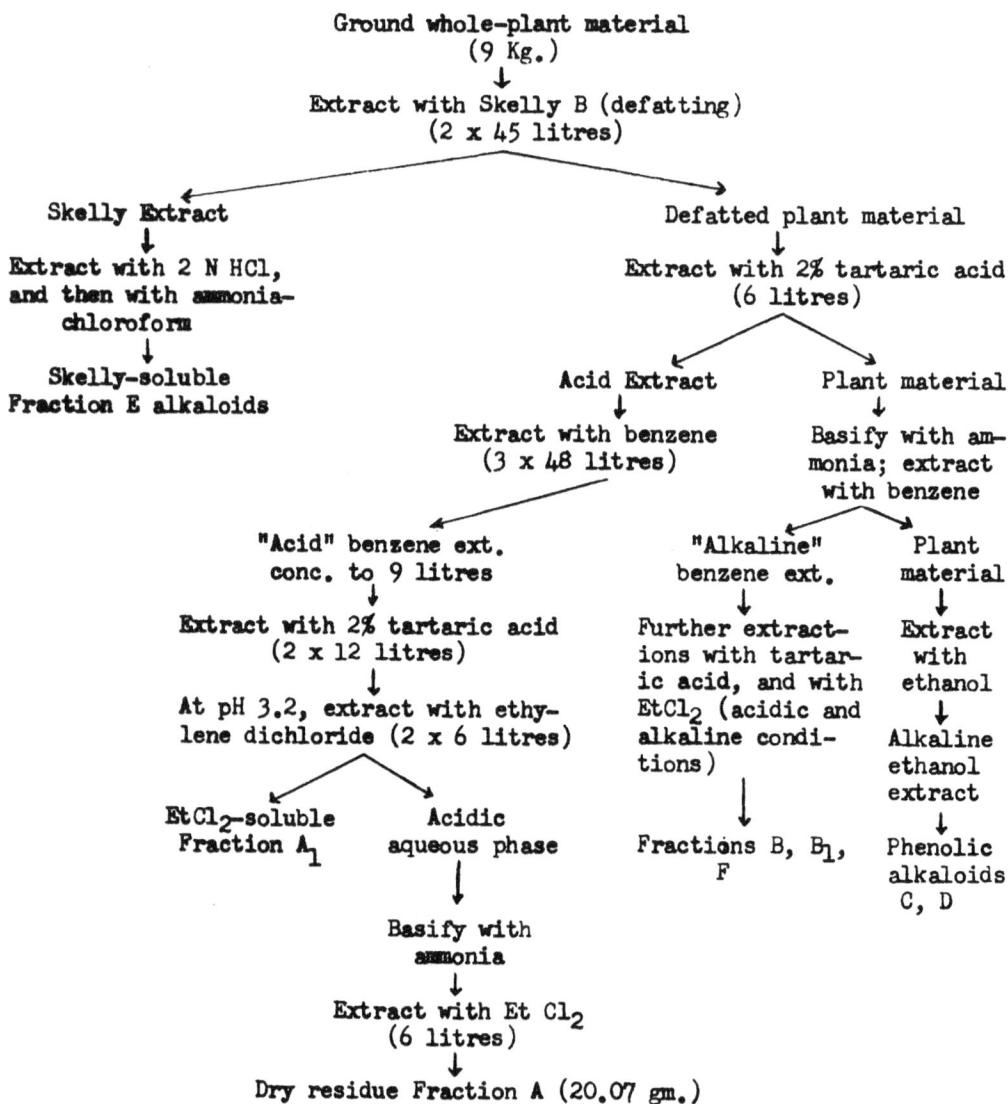

Ground whole-plant material
(9 Kg.)
↓
Extract with Skelly B (defatting)
(2 x 45 litres)

Skelly Extract
↓
Extract with 2 N HCl,
and then with ammonia-
chloroform
↓
Skelly-soluble
Fraction E alkaloids

Defatted plant material
↓
Extract with 2% tartaric acid
(6 litres)

Acid Extract
↓
Extract with benzene
(3 x 48 litres)

Plant material
↓
Basify with am-
monia; extract
with benzene

"Acid" benzene ext.
conc. to 9 litres
↓
Extract with 2% tartaric acid
(2 x 12 litres)
↓
At pH 3.2, extract with ethylene dichloride (2 x 6 litres)

"Alkaline"
benzene ext.
↓
Further extract-
ions with tartar-
ic acid, and with
EtCl$_2$ (acidic and
alkaline condi-
tions)
↓
Fractions B, B$_1$,
F

Plant
material
↓
Extract
with
ethanol
↓
Alkaline
ethanol
extract
↓
Phenolic
alkaloids
C, D

EtCl$_2$-soluble
Fraction A$_1$

Acidic
aqueous phase
↓
Basify with
ammonia
↓
Extract with Et Cl$_2$
(6 litres)
↓
Dry residue Fraction A (20.07 gm.)

Figure 7. Svoboda Extraction Scheme
for Vinca Alkaloids

form 1:3 (Fractions 30 - 43); chloroform (Fractions 44 - 57); and chloroform:methanol 19:1 (Fractions 58 - 67). Fraction 1 is 1500 litres, and succeeding fractions are 150 litres (56).

In these re-chromatographed fractions, <u>vinrosidine</u> is isolated from Fractions 33 - 45, and <u>vincristine</u> is isolated from Fractions 33 - 42. The isolation of these two alkaloids from these column fractions is achieved by subjecting each fraction to gradient pH extractions.

<u>Gradient pH Extractions</u>: - The eluted crude fraction (10 gm.) is dissolved in 500 ml. of benzene, and any insoluble material is removed by filtration. The benzene solution is then extracted into 500 ml. of 0.1 M citric acid by steam distillation under reduced pressure. This steam distillation procedure (called "acid water run-down" technique) is used in order to obviate emulsion-formation. After removal of all of the benzene, any insoluble material is filtered off. The acidic solution (pH 2.75 - 2.85) is extracted with 500 ml. of benzene. The aqueous phase is then adjusted with ammonia, successively, to pH 3.4, 3.9, 4.4, 4.9, 5.4, 5.9, 6.4, and 7.5, and at each pH level the aqueous solution is extracted with 500 ml. of benzene. The benzene extract in each case is dried over sodium sulphate and evaporated to dryness. Both <u>vinrosidine</u> and <u>vincristine</u> are extracted in this manner from the acidic solution at the pH levels 4.9 - 6.4 (56).

Vinrosidine is crystallized from methanol or ethanol, vincristine from methanol, and vincristine sulphate from ethanol. In the case of those fractions containing approximately 1:1 mixture of these two alkaloids, the mixture (residue obtained from the gradient pH extraction) is dissolved in hot ethanol, and chilling the solution produces vinrosidine which crystallizes out. The mother liquor is then evaporated to dryness and the residue is dissolved in hot methanol and on chilling the methanolic solution vincristine is crystallized out.

(d) Action and Medicinal Uses

A prominent biological activity of vinblastine is that it produces leukopenia in rats and other animals (34), and it interferes with deoxyribonucleic acid synthesis in animal thymus cell system in vitro and in vivo (49). Vinblastine, vincristine, vinleurosine, and vinrosidine have been shown to be effective in the treatment of neoplastic disease in animals and in man (34). Among the Vinca alkaloids, vinblastine and vincristine have been studied more extensively in their clinical use than the other alkaloids, and their effectiveness and usefulness in the treatment of Hodgkin's disease, lymphosarcoma, and acute leukemia have been well demonstrated (23, 34). In spite of the small difference in their chemical structures, these two alkaloids differ greatly in their antitumour activity spectrum and in their toxicity (34). Both alkaloids are effective in producing partial or complete tumour regression or remission in patients with Hodgkin's disease, but vincristine is the more toxic of the two. They are less effective in the treatment of lymphosarcoma than in the treatment of Hodgkin's disease. Vincristine is therefore recommended for induction of such remissions but not for remission maintenance (23, 34). Vincristine is much less active in adults with acute leukemia and inactive in chronic leukemia. Vinblastine is not effective in the treatment of acute leukemia. Vinblastine produces leukopenia (its major toxic effect) more regularly than vincristine. Both alkaloids increase the mitotic index in the bone marrow due to arrest of cells in the metaphase stage of mitosis.

Some of the monomeric Vinca alkaloids (e.g. catharanthine and vindo-linine) have been shown to possess diuretic properties (25a).

Pharmaceutical specialty products are Velban[R] (Velbe[R]) containing vinblastine sulphate, and Oncovin[R] containing vincristine sulphate.

8.6 References

1. Abe, M. – J. Agr. Chem. Soc. Japan, 22:2 (1948); Chem. Abstr., 46: 3218 (1952).

2. Abe, M. – Ann. Report of Takeda Res. Lab. (Osaka, Japan), 22:116-124 (1963).

3. Abe, M., et al. – J. Agr. Chem. Soc. Japan, 25:458 (1952); 28:44-47 (1954); 34:246-249 (1960); 34:360-371 (1960).

4. Abe, M., et al. – Bull. Agr. Chem. Soc. Japan, 19:92 (1955).

5. Agurell, S., & Ramstad, E. – Lloydia, 25:67-77 (1962).

6. Arcamone, F., Chain, E. B., Ferretti, A., Minghetti, A., Penella, P., Tonolo, A., & Vero, L. – Proc. Roy. Soc. (London), B155:26-54 (1961).

7. Baxter, R. M., Kandel, S. I., & Okany, A. – Chem. and Ind., p. 266 (1960); Nature, 185:241 (1960).

8. Baxter, R. M., Kandel, S. I., & Okany, A. – Chem. and Ind., p. 1453 (1961); Can. J. Chem., 42:2936-2938 (1964).

9. Baxter, R. M., Kandel, S. I., & Okany, A. – J. Am. Chem. Soc., 84: 2997-2999 (1962).

10. Bhattacharji, S., Birch, A. J., Brack, A., Hofmann, A., Kobel, H., Smith, D. C. C., Smith, H., & Winter, J. – J. Chem. Soc., pp. 421-425 (1962).

11. Brady, L. R. – Lloydia, 25:1-36 (1962).

12. Brindle, H., Carless, J. E., and Woodhead, H. B. – J. Pharm. Pharmacol., 3:793-813 (1951).

13. Charterjee, A., Pakrashi, S. C., & Werner, G. – in L. Zechmeister (ed.): Progress in the Chemistry of Organic Natural Products, Vol. 13, pp. 346-407, Springer, Vienna, 1956.

14. Clarke, E. G. C. - J. Pharm. Pharmacol., 9:187-192 (1957).

15. Cromwell, B. T. - in K. Paech & M. V. Tracey (eds.): Modern Methods of Plant Analysis, Vol. IV, (a) pp. 411-412; (b) pp. 480-487, Springer Berlin, 1955.

16. Dudley, H. W., & Moir, J. C. - Brit. Med. J., 1:520 (1935).

17. Dutt, A., et al. - Indian J. Pharm., 9:54-57 (1947).

18. Dwyer, J. D. - Lloydia, 27:282-285 (1964).

19. El. Ridi, M. S., and Khalifa, K. - J. Pharm. Pharmacol., 4:190-196 (1952).

20. Farnsworth, N. R. - Lloydia, 24:105-138 (1961).

21. Floss, H. G., Mothes, U., & Gunther, H. - Z. Naturforsch., 19b:784-788 (1964).

22. Foster, G. E., Macdonald, J., & Jones, T. S. G. - J. Pharm. Pharmacol., 1:802-812 (1949).

23. Frei, E., III - Lloydia, 27:364-367 (1964).

24. Gorman, M., & Neuss, N. - Lloydia, 27:393-396 (1964).

25. Gorman, M., Neuss, N., and Cone, N. J. - J. Am. Chem. Soc., 87:93-99 (1965).

25a. Gorman, M., Tust, R. H., Svoboda, G. H., and Le Men, J. - Lloydia, 27:214-219 (1964).

26. Gröger, D., & Mothes, K. - Pharmazie, 11:323 (1956).

27. Gröger, D., et al. - Z. Naturforsch., 14b:355-358 (1959).

28. Gröger, D., et al. - Z. Naturforsch., 15b:141-143 (1960).

29. Gröger, D., Tyler, V. E. Jr., & Dusenberry, J. E. - Lloydia, 24:97-102 (1961).

30. Hoffer, A. - Clin. Pharmacol. & Therap., 6:183-255 (1965).

31. Hofmann, A. - Helv. Chim. Acta, 40:1358-1373 (1957).

32. Holmes, H. L. - in R. H. F. Manske & H. L. Holmes (eds.): The Alkaloids, Vol. I, pp. 375-500, and Vol. II, pp. 513-551, Academic Press, New York, 1950-1952.

33. Jindra, A., and Pohorski, J. - J. Pharm. Pharmacol., 3:344-350 (1951).

34. Johnson, I. S., Armstrong, J. G., Gorman, M., and Burnett, J. P. Jr., - Cancer Res., 23:1390-1427 (1963).

35. Klohs, M. W., et al. - J. Am. Chem. Soc., 75:4867 (1953).

36. Klohs, M. W., et al. - J. Am. Chem. Soc., 76:2843 (1954).

37. Klohs, M. W., et al. - J. Am. Chem. Soc., 77:2241-2243 (1955).

38. Lucas, R. A. - in G. P. Ellis & G. B. West (eds.): Progress in Medicinal Chemistry, Vol. 3, pp. 146-186, Butterworth, London, 1963.

39. Marion, L. - in R. H. F. Manske & H. L. Holmes (eds.): The Alkaloids, Vol. II, (a) pp. 375-393, (b) pp. 438-440, Academic Press, New York, 1952.

40. Mothes, K., Weygand, F., Gröger, D., & Grisebach, H. - Z. Naturforsch., 13b:41-44 (1958).

41. Müller, R., Schlittler, E., & Bein, H. J. - Experientia, 8:338 (1952).

42. Neuss, N., Boaz, H. E., & Forbes, J. W. - J. Am. Chem. Soc., 75:4870 (1953); 77:4087-4090 (1955).

43. Neuss, N., Gorman, M., Boaz, H. E., & Cone, N. J. - J. Am. Chem. Soc., 84:1509-1510 (1962).

44. Noble, R. L., Beer, C. T., & Cutts, J. H. - Ann. N. Y. Acad. Sci., 76:882-894 (1958).

45. Plieninger, H., et al. - Ann., 642:214-224 (1961).

46. Plieninger, H., Fischer, R., & Liede, V. - Ann., 672:223-231 (1964).

47. Pöhm, M. - Arch. d. Pharm., 291:468-480 (1958).

48. Quirin, M., Lévy, J. and Le Men, J. - Ann. Pharm. Franc., 23:93-98 (1965).

49. Richards, J. F., & Beer, C. T. - Lloydia, 27:346-351 (1964).

50. Saxton, J. E. - in R. H. F. Manske (ed.): The Alkaloids, Vol. VII, (a) pp. 9-34, (b) p. 146, Academic Press, New York, 1960.

51. Schlientz, W., Brunner, R., Stadler, P. A., Frey, A. J., Ott, H., & Hofmann, A. - Helv. Chim. Acta, 47:1921-1933 (1964).

52. Spilsbury, J. F., & Wilkinson, S. - J. Chem. Soc., p. 2085 (1961).

52a. Stauffacher, D., and Tscherter, H. - Helv. Chim. Acta, 47:2186-2194 (1964).

53. Stoll, A. - Helv. Chim. Acta, 28:1283-1308 (1945).

54. Stoll, A. - Chem. Rev., 47:197-218 (1950).

55. Stoll, A., et al. - Helv. Chim. Acta, 37:1815-1825 (1954).

56. Svoboda, G. H. - Lloydia, 24:173-178 (1961).

57. Svoboda, G. H., Johnson, I. S., Gorman, M., & Neuss, N. - J. Pharm. Sci., 51:707-720 (1962).

58. Svoboda, G. H., Neuss, N., & Gorman, M. - J. Am. Pharm. Assoc. Sci. Ed., 48:659-666 (1959).

59. Taber, W. A., & Vining, L. C. - Can. J. Microbiol., 3:55-60 (1957).

60. Taber, W. A., & Vining, L. C. - Chem. and Ind., p. 1218 (1959).

61. Taylor, E. H. - Am. J. Pharm. Educ., 28:205-210 (1964).

62. Taylor, E. H., & Ramstad, E. - J. Pharm. Sci., 50:681-683 (1961).

63. Weygand, F., & Floss, H. G. - Angew. Chem., Intern. Ed., 2:243-247 (1963).

64. Woodson, R. E., Youngken, H. W., Schlittler, E., & Schneider, J. A. - Rauwolfia: Botany, Pharmacognosy, Chemistry, and Pharmacology, Little, Brown, Boston-Toronto, 1957.

CHAPTER 9

STEROIDAL ALKALOIDS

9.1   Introduction

The steroidal alkaloids of medicinal significance are the Veratrum alkaloids. The ring systems in the molecules of many of the Veratrum alkaloids do not follow exactly the pattern of ring systems in the regular steroid molecule such as in cholesterol, the aglycones of cardiac glycosides, etc. As may be seen from the structures shown in the next section, in the alkamine portion of the ester alkaloids of Veratrum (protoverine, veracevine, germine, etc.) and in the alkamine aglycones of some of the glycosidic Veratrum alkaloids (veratramine, etc.), ring C is a 5-membered ring and ring D is a 6-membered ring, which is the reverse of the pattern in the regular steroid molecule.

9.2   Veratrum Alkaloids

(a) Plant Sources

More than a score of steroidal alkaloids have been isolated and chemically characterized from various Veratrum species (Fam. Liliaceae), from Sabadilla, and from Zygadenus species. A number of these have proved to be pharmacologically and therapeutically useful. Some of the better known plant species bearing these Veratrum alkaloids are listed in Table 10.

(b) Types of Principal Alkaloids

Three groups may be distinguished from among the alkaloids occurring in these species:

1. Alkamines - secondary and tertiary amines.

2. Glycosidic alkaloids - an alkamine is linked through its hydroxyl group at C-3 in a glycosidic linkage to a sugar.

3. Ester alkaloids - an alkamine is linked, through its several alcoholic hydroxyl groups by ester linkages, to various acids.

Table 10. Some Plant Sources of
the Veratrum Alkaloids

Plant species	Common names	Plant parts bearing the alkaloids
Veratrum viride Aiton	Green Hellebore	Root & rhizome
Veratrum album L. (V. escholtzii Gray)	White Hellebore	Root & rhizome
Schoenocaulon officinale Gray (Veratrum sabadilla Retz (Sabadilla officinalis Brand et Ratzeb)	Sabadilla; Cevadilla.	Seeds
Zygadenus venenosus Watson		Root and aerial parts

Also, from the point of view of structure types, the majority of these Veratrum alkaloids may be classified into two major groups, the "Ceveratrum group" and the "Jerveratrum group" (6), although a few of the alkaloids cannot be classified with either one of these two groups in structure type.

(i) Ceveratrum Group - represented by the structures of protovera-

- 148 -

trines, zygadenine, germine, cevadine, veratridine, etc. (the ester alkaloids) in which the alkamine portion has many (usually 6 to 9) hydroxyl
groups linked to various acids, and there is an oxygen bridge between C-4
and C-9.

(ii) Jerveratrum Group - represented by the structures of veratramine, jervine, isorubijervine, etc., in which the ring systems beyond C-17
(the last two of the six rings) are different from those in the Ceveratrum
group, there is absence of the oxygen bridge between C-4 and C-9 and absence
of numerous hydroxyl groups, and presence of a double-bond between C-5 and
C-6.

The Zygadenus species and the Schoenocaulon species appear to have
only the Ceveratrum type and no Jerveratrum type of alkaloids, while the
various Veratrum species appear to have both of these two structure types of
alkaloids (6).

Table 11. Some Alkamines of Veratrum
and Zygadenus Species

Alkamine	Plant species
Germine	Zygadenus venenosus
Isorubijervine	Veratrum viride Aiton, Veratrum album L.
Jervine	Veratrum album L.
Rubijervine	Veratrum album L.
Veratramine	Veratrum viride, Veratrum grandiflorum O. Loes.
Zygadenine	Zygadenus venenosus Watson, Zygadenus indermedius Rydb.

Table 12. Glycosidic Alkaloids
of Veratrum Species

Glycoalkaloid	Product of Hydrolysis with 2% HCl		Plant species
	Alkamine	Sugar	
Isorubijervosine	Isorubijervine	D-glucose	V. album
Pseudojervine	Isojervine	D-glucose	V. album
Veratrosine	Veratramine	D-glucose	V. viride

In Tables 11, 12 and 13 (1, 3, 4, 5, 7) are listed some of the alkaloids of these three groups which have been fairly well characterized chemically. Medicinally the most widely used ones among these (as purified alkaloids) are protoveratrines A and B. The mixture of these two ester alkaloids was formerly referred to as "protoveratrine".

The term "veratrine" is sometimes used to mean a mixture of alkaloids obtained from the seeds of Schoenocaulon officinale, consisting mainly of the ester alkaloids veratridine and cevadine.

The term "alkavervir" is used to designate a mixture of alkaloids (alcohol soluble and practically water-insoluble) extracted by selective acidic and basic precipitation from Veratrum viride and standardized biologically for hypotensive effect.

The terms "cryptenamine" acetate or tannate refer to a mixture of these salts of the alkaloids extracted from Veratrum viride.

Table 13. Ester Alkaloids of Veratrum
and Related Species

Ester alkaloid	Products of Alkaline Hydrolysis		Plant species
	Alkamine	Acids	
Protoveratrine A	Protoverine	2(acetic); 2-methylbutyric; 2-methyl-2-hydroxybutyric.	V. album
Protoveratrine B	Protoverine	2(acetic); 2-methylbutyric; 2-methyl-2,3-dihydroxybutyric.	V. album
Cevadine	Veracevine	angelic (tiglic)	S. officinale
Cevacine	Veracevine	acetic	S. officinale
Veratridine	Veracevine	veratric	S. officinale
Germidine	Germine	acetic; 2-methylbutyric	V. viride
Germerine	Germine	2-methylbutyric; 2-methyl-2-hydroxybutyric	V. viride; V. album
Germitrine	Germine	acetic; 2-methylbutyric; 2-methyl-2-hydroxybutyric	V. viride; V. album
Protoveratridine	Germine	2-methylbutyric	Z. venenosus
Zygacine	Zygadenine	acetic	Z. venenosus
Vanilloyl-zygadenine	Zygadenine	vanillic	Z. venenosus
Veratroyl-zygadenine	Zygadenine	veratric	Z. venenosus

^{27}Me H

^{25} ^{24} ^{26} ^{23}

N ^{21} ^{18} ^{22} Me

H ^{20} OH

^{19} Me ^{11} ^{12} ^{13} ^{17}

^{9} ^{14} ^{16} OH

^{1} ^{10} ^{8} ^{15} OH

^{2} O OH OR$_4$

^{3} ^{5} ^{7} OR$_3$

R$_1$O ^{6}

OH OR$_2$

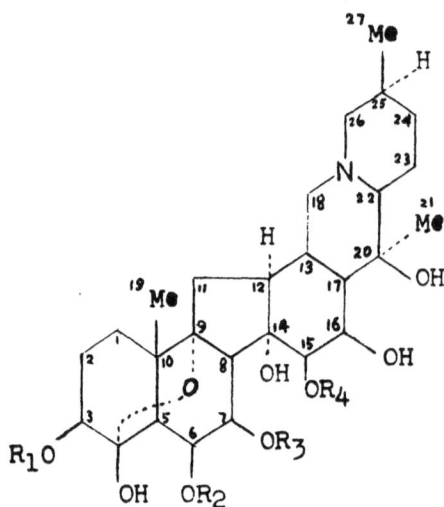

	R_1	R_2	R_3	R_4
Protoverine	H	H	H	H
Protoveratrine A	2-hydroxy-2-methylbutyryl	acetyl	acetyl	2-methyl-butyryl
Protoveratrine B	2,3-dihydroxy-2-methylbutyryl	acetyl	acetyl	2-methyl-butyryl
Zygadenine	H	H replaces OR_2	H replaces OR_3	H
Zygacine	acetyl	H replaces OR_2	H replaces OR_3	H
Germine	H	H replaces OR_2	H	H
Germidine	acetyl	H replaces OR_2	H	2-methyl-butyryl

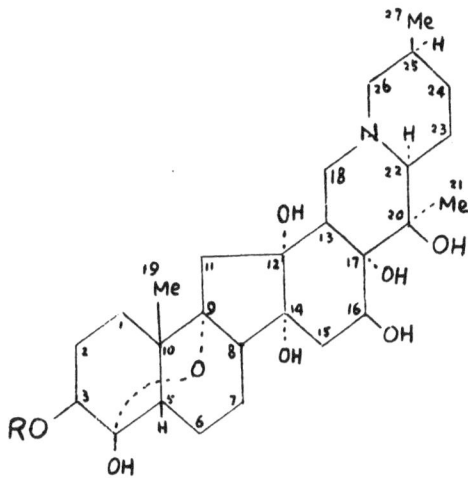

$$R =$$

Veracevine	H
Veratridine	veratryl
Cevadine	tiglyl
Cevacine	acetyl

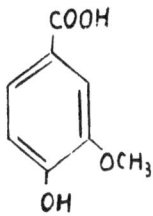

Vanillic Acid

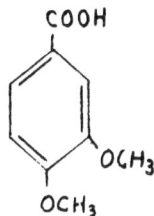

Veratric Acid

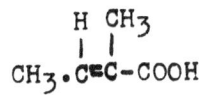

$$CH_3 \cdot C \equiv C - COOH$$

Angelic Acid
(Tiglic Acid)

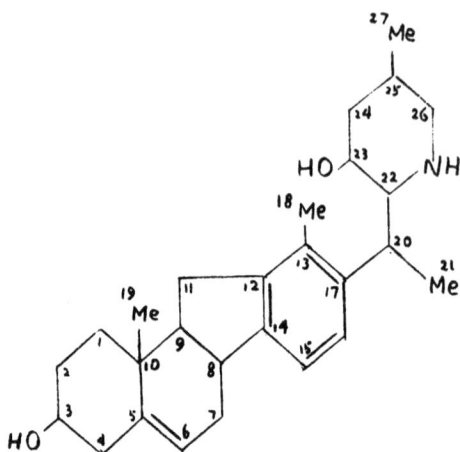

Veratramine

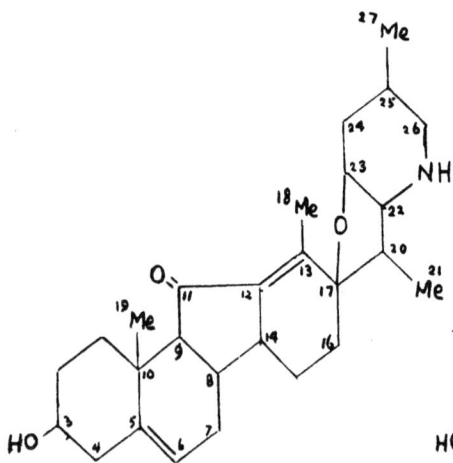

Jervine

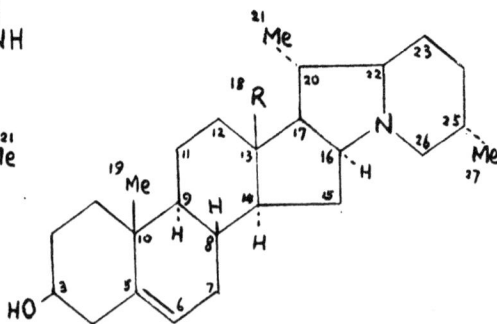

Isorubijervine, R = CH₂OH

(c) Isolation of Protoveratrines

Protoveratrines A and B were not separated until a few years ago. These were usually extracted together and referred to as "protoveratrine". In the procedure of Craig and Jacobs (2), the plant material (V. album), in 2-Kg. batches, is first extracted with benzene and ammonia. The alkaloids are then purified by extraction into acetic acid, re-extracted into benzene, and, after removal of solvent, dissolved in ether from which crystalline powder of the crude protoveratrines separates out and is recrystallized from alcohol-acetic acid upon alkalinization of the solution. About 9 gm. of crude protoveratrine powder may be obtained from 8 Kg. of V. album roots. This procedure may be summarized in outline as shown in Figure 8.

The A and B components may be separated by counter-current distribution of the "protoveratrine" between benzene and acetate buffered at pH 5.5, followed by column chromatography on acid aluminum oxide (7).

Protoveratrines A and B are soluble in chloroform, slightly soluble in ether, and practically insoluble in petroleum ether or water. They are fairly stable in acid medium (pH 4 - 6), but they rapidly decompose in alkaline solution.

(d) Pharmacological Action and Uses

The veratrum alkaloids possess hypotensive action through reflex inhibition of presso receptors in the heart and carotid sinus. They also have emetic action and may cause nausea in hypotensive dosage. They are, therefore, often used in combination with Rauwolfia alkaloids for hypotensive effects. Veratrum plant extracts and the alkaloids (mainly protovera-

2 Kg. plant material stirred over-night
with 7 litres of benzene, 1 litre of
water and 100 ml. of ammonia

↓

Filter and re-extract the solid
with two 5-litre portions of benzene

↓

Benzene extract derived from four 2-Kg.
batches of plant material is concentrated to
3 litres volume

↓

Extract with one-litre portions of 5%
acetic acid (6 x 1 litres)

↓

Acid extract made alkaline with 25% NaOH and
shaken with benzene

↓

Benzene ext. repeatedly washed with water

↓

Benzene extract evaporated to dryness

↓

Dissolve residue in 500 ml. dry ether

↓

Standing 18 hours

↓

Crystalline powder of crude protoveratrine

↓

Suspend in 10 parts of 95% ethanol and add
excess of acetic acid

↓

Add slight excess of ammonia

↓

Crystalline powder of protoveratrine

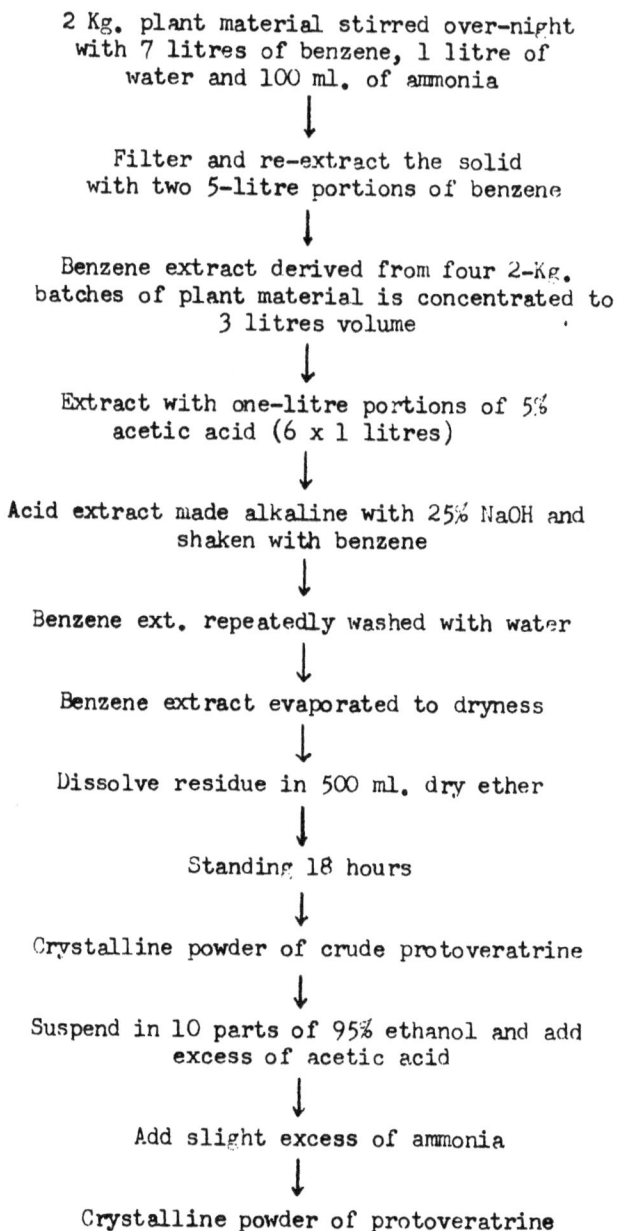

Figure 8.   Procedure Outline for Isolation of Protoveratrines

trines A and B) have been used in treatment of toxaemia of pregnancy and for management of hypertensive crises and eclampsia.

(e) Medicinal Preparations (Some Pharmaceutical Specialties)

(a) Vertavis[R] (Veratrum viride rhizome and root material)

(b) Those containing Veratrum alkaloidal extracts:

Veratrite[R] (with other medicinal agents); Vergitryl[R];

Veriloid[R]; Unitensin[R]; Rauvera[R] (with Rauwolfia extract);

Veraflex[R] (with other medicinal agents).

(c) Those containing purified alkaloids:

Protoveratrines A and B (N.N.D.); Protalba[R] (protoveratrine A); Provell Maleate[R] (protoveratrines A and B); Veralba[R] (Protoveratrines with Rauwolfia); Veralba-R[R] (Protoveratrines A and B with reserpine).

9.3 References

1. Barton, D. H. R., et al. - Experientia, 10:81 (1954).

2. Craig, L. C., & Jacobs, W. A. - J. Biol. Chem., 143:427-432 (1942).

3. Kupchan, S. M., Ayres, C. I., Neeman, M., Hensler, R. H., Masamune, T., & Rajagopalan, S. - J. Am. Chem. Soc., 82:2242-2258 (1960).

4. Kupchan, S. M., & Deliwala, C. V. - J. Am. Chem. Soc., 74:2382-2383; 74:3202 (1952).

5. Kupchan, S. M., Johnson, W. S., & Rajagopalan, S. - J. Am. Chem. Soc., 80:1769 (1958).

6. Kupchan, S. M., Zimmerman, J. H., & Alfonso, A. - Lloydia, $\underline{24}$:1-26 (1961).

7. Prelog, V., & Jeger, O. - in R. H. F. Manske & H. L. Holmes (eds.): The Alkaloids, Vol. 1II, pp. 270-312, and Vol. VII, pp. 363-417, Academic Press, New York, 1953 and 1960.

CHAPTER 10

ALKALOIDS OF THE IMIDAZOLE GROUP

10.1 Pilocarpus Alkaloids

(a) Plant Sources

A number of __Pilocarpus__ species (Fam. Rutaceae) contain several alkaloids having an imidazole ring in the molecule. The following are some of
the better known species:

Plant species	Common name
__Pilocarpus jaborandi__ Holmes	Pernambuco Jaborandi
__Pilocarpus pennatifolius__ Lemaire	Paraquay Jaborandi
__Pilocarpus microphyllus__ Staph	Maranham Jaborandi
__Pilocarpus racemosus__ Vahl.	Guadeloupe Jaborandi

The alkaloids occur in these species largely in the leaves, and the
major ones are pilocarpine and its stereoisomer isopilocarpine. Pilocarpine
constitutes 0.2 - 0.7% of the dried leaf material. The proportion of isopilocarpine appears to vary considerably, ranging from 5 to 75% of the total
alkaloids. In most samples reported in the literature isopilocarpine seems
to constitute about one-third to one-half of the quantity of pilocarpine.

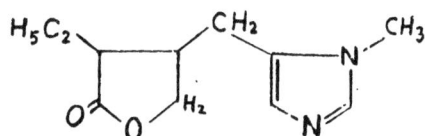

Pilocarpine
Isopilocarpine

(b) Properties and Tests

Pilocarpine and isopilocarpine are oily liquids, soluble in water, alcohol or chloroform, fairly soluble in benzene, and nearly insoluble in ether or petroleum ether. Pilocarpine exhibits an absorption maximum at 263 mµ.

Pilocarpine behaves as a monoacidic base. It gives precipitates with silicotungstic and phosphomolybdic acids, gold and platinic halides, Wagner's, Mayer's, and Hager's reagents (see pages 9 and 10). Some of these precipitates serve as aid to identification of the alkaloid. The lactone ring is opened by caustic alkali which forms salts with the resulting acid. This opening of the lactone ring destroys the physiological activity of pilocarpine. The lactone ring is unaffected by ammonia or alkali carbonates. Permanganate oxidation of pilocarpine destroys the imidazole ring and yields ammonia, methylamine, pilopic and homopilopic acids, and other products.

Helch's Colour Test: - A small quantity (0.5 ml.) of pilocarpine solution is treated with a small crystal of potassium dichromate and 1 - 2 ml. of chloroform, followed by addition of 1 ml. of 3% hydrogen peroxide and shaking the mixture. The chloroform layer becomes blue.

Ekkert's Colour Test: - To 1 ml. of 1% solution of pilocarpine hydrochloride are added 1 ml. of 2% sodium nitroprusside solution and 1 ml. of 1 N sodium hydroxide solution. After a few minutes standing, the mixture is acidified with dilute hydrochloric acid, and thereupon a wine or red colour appears. This test is also given by isopilocarpine hydrochloride. When a few drops of 0.1 N sodium thiosulphate solution are then added, the colour changes to green(2).

Quantitative estimation of pilocarpine may be carried out by titration of the isolated alkaloid with standard acid, using bromophenol blue, methyl red, or methyl orange as indicator.

Isomerism

Pilocarpine and isopilocarpine are stereoisomers, having the stereochemical difference in the lactone portion of the molecule (1). This conclusion is based on experimental evidence which shows that the isomerism of the two alkaloids still persists when the imidazole ring is destroyed under mild conditions. For example, ozonolysis of the methylated derivatives of the two alkaloids yields different products.

Pilocarpine and isopilocarpine may be separated only with great difficulty through repeated crystallization of their different salts. They also readily undergo interconversion to each other. Pilocarpine is isomerized to isopilocarpine when heated in aqueous or alcoholic sodium hydroxide solution, or heated with fuming nitric acid. Prolonged heating of pilocarpine hydrochloride solution causes isomerization to isopilocarpine hydrochloride.

(c) Pharmacological Action and Uses

Pilocarpine stimulates the organs innervated by post-ganglionic cholinergic fibres to produce parasympathomimetic effects. It therefore possesses diaphoretic and miotic actions and is used to reduce intra-ocular pressure in glaucoma.

Medicinal Preparations

Pilocarpine Nitrate U.S.P. XVII, B.P.C. 1963

Pilocarpine Hydrochloride U.S.P.

10.2 References

1. Battersby, A. R., & Openshaw, H. T. - in R. H. F. Manske and R. L. Holmes (eds.): The Alkaloids, Vol. III, pp. 206-228, Academic Press, New York, 1953.

2. Cromwell, B. T. - in K. Paech & M. V. Tracey (eds.): Modern Methods of Plant Analysis, Vol. 4, pp. 471-472, Springer, Berlin, 1955.

CHAPTER 11

ALKALOIDS OF THE PHENETHYLAMINE GROUP

The principal alkaloidal amines which are phenethylamine derivatives
and which have significant pharmacological properties are the Ephedra alkaloids and the Peyote alkaloids.

11.1 Ephedra Alkaloids

(a) Plant Sources

Several Ephedra species (Family Gnetaceae, Subdivision Gymnospermae)
constitute the Chinese drug Ma Huang which has been used as medicine in
China for centuries. These species contain several alkaloids which are β-
phenethylamine derivatives, of which the medicinally important ones are D-
(-)-ephedrine and its stereoisomer L-(+)-pseudophedrine (also known as (+)-
ψ-ephedrine). These are now produced for medicinal use by chemical synthesis. D-(+)-ephedrine and L-(-)-pseudoephedrine have been made syntheti-
cally but have not been found to occur in these plants (4b).

The species which have been more intensely investigated chemically
and pharmacologically and found to have high contents of ephedrine (1 to 3%)
are: Ephedra equisetina Bunge, E. sinica Stapf, E. distachya L., E. inter-
media Schrenk et C. A. Mey, and E. nebrodensis Tineo. These species are
mostly found in China and India, but have also been found in Europe. The
American species Ephedra americana Humb. et Bonpl. has been reported to have
a rather low (0.38%) ephedrine content, and a number of other American species of Ephedra have been reported to be devoid of alkaloids (4b). In the
alkaloid-bearing species, the alkaloids occur in the practically leafless

- 163 -

slender stems of the plants, which are all low-growing (two to three feet) shrubs. The ratio of ephedrine to pseudoephedrine varies widely in different plants. The more mature plants are said to have higher alkaloidal contents, especially in the fall.

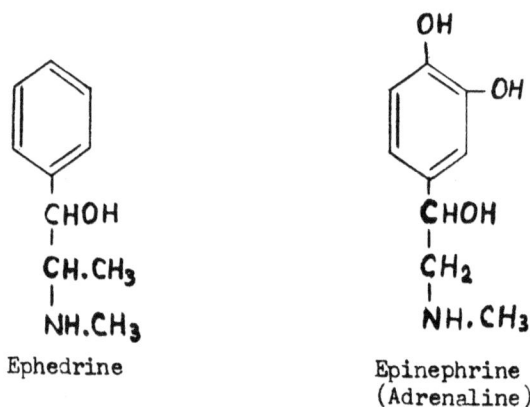

Ephedrine

Epinephrine
(Adrenaline)

As may be seen from the structure formulae, ephedrine is closely related to the animal hormone epinephrine. Indeed they rather resemble each other qualitatively in their pharmacological (adrenergic) properties in which epinephrine is much the more potent.

Ephedrine and pseudoephedrine may be extracted from the plant material by general procedures for alkaloid extraction, through successive benzene and dilute hydrochloric acid extractions. Ephedrine may be separated from pseudoephedrine by means of their oxalates, ephedrine oxalate being much less soluble in cold water than pseudoephedrine oxalate. Chloroform is not a suitable solvent for the extraction of these alkaloids as ephedrine in chloroform solution on evaporation yields ephedrine hydrochloride and aldehyde.

(b) Extraction and Properties

Ephedrine and pseudoephedrine are rather unusually stable among alkaloids. They can be heated at 100° C. for several hours without decomposition. However, ephedrine solutions are unstable in light in the presence of
oxygen.

Ephedrine hydrochloride when heated with 25% hydrochloric acid is
partially converted to pseudoephedrine (1, 4b). This conversion is rever-
sible, and an equilibrium is established.

N-acetylation of ephedrine (and of pseudoephedrine) may be effected
with acetic anhydride at 70° C. (5). Refluxing N-acetyl-(-)-ephedrine with
5% hydrochloric acid yields a mixture of (-)-ephedrine and (+)-pseudoephe-
drine.

Heating (-)-ephedrine hydrochloride yields some 52% of O-acetyl-(+)-
pseudoephedrine. N-acetyl-($\pm$)-ephedrine on standing for several days in
acetone solution containing hydrochloric acid gives O-acetyl-($\pm$)-ephedrine,
which in the presence of alkali rearranges back to the N-acetyl-($\pm$)-ephed-
rine (5). ($\pm$)-Ephedrine is also known as racephedrine, and as ephetonin.

(c) Pharmacological Action and Uses

Ephedrine exerts sympathomimetic actions similar to epinephrine
(ephedrine is effective by oral administration, whereas epinephrine is not).
In addition, ephedrine exerts excitatory action on the central nervous system and produces marked effects on skeletal muscles. Ephedrine, in the form
of its salts, is widely used in symptomatic relief of asthma and certain
other allergic disorders as bronchodilator and as nasal decongestant. It
is also a useful adjunct in the treatment of myasthenia gravis. Pseudo-

ephedrine possesses pharmacological properties similar to those of ephedrine. Pseudoephedrine is said to be less toxic in certain respects than ephedrine. The relative pressor activities (6) of ephedrine and its isomers are as follows:

	Relative Pressor Activity*
D-(-)-ephedrine	36
D-(+)-ephedrine	11
D-(±)-ephedrine	26
L-(+)-pseudoephedrine	7
L-(-)-pseudoephedrine	1

*Adapted from C. O. Wilson and O. Gisvold - Textbook of Organic Medicinal and Pharmaceutical Chemistry, ed. 4, J. B. Lippincott Co., Philadelphia, 1962.

(d) Medicinal Preparations

Ephedrine B.P.C. 1963

Elixir of Ephedrine B.P.C.

Nasal Drops of Ephedrine B.P.C.

Ephedrine Hydrochloride N.F. XII, B.P. 1963

Ephedrine Sulphate U.S.P. XVII, N.F. XII

Racephedrine Hydrochloride N.F. XII

Pharmaceutical Specialties

There are many pharmaceutical specialties containing ephedrine salt as the principal active ingredient or ephedrine in combination with certain barbiturates and/or other medicinal agents (7). The following are a few examples:

Bena-Fedrin[R]; Calcidrine[R]; Dainite[R]; Extosen[R]; I-Sedrin[R]; Quadrinal[R]; Tedral[R]; Sudafed[R] (pseudoephedrine hydrochloride).

11.2 Alkaloids of Peyote

(a) Plant Sources

Peyote (Pellote, Mescal Buttons), which has been used in connection with religious rituals of certain Indian tribes of Mexico and south-western United States, consists of the tops of the cactus **Lophophora williamsii** (Lem) Coulter (formerly known as **Anhalonium lewinii** Hennings), Family Cactaceae. This species is also known as **Echinocactus williamsii** Lem. It contains several β-phenethylamine derivatives of which the principal one is mescaline. Peyote contains up to 6% of mescaline, which has also been found in other species of cacti.

(b) Properties

Mescaline (4a) is 3,4,5-trimethoxy-phenethylamine. Mescaline (the free base) is a strongly alkaline oil (b.p. 180° C.) or it may be in crystalline form (m.p. 35 - 36° C.). It is soluble in water, alcohol, or chloroform, but only slightly soluble in ether. It may be isolated from the plant material as mescaline sulphate which is insoluble in alcohol, only slightly soluble in cold water, but very soluble in hot water. Mescaline also forms hydrochloride, picrate, aurichloride and other salts.

Heating of mescaline with acetic acid at 170 - 175° C. yields N-acetylmescaline, which also occurs in the plant material (mescal buttons) (4a).

(c) Pharmacological Action

Mescaline is a hallucinogenic agent which induces colour visions and depression of some functions (especially errors in the estimation of time). Large doses depress respiration. It is of pharmacological interest because

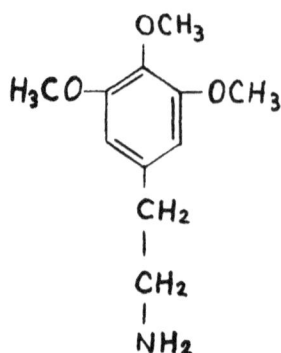

$$
\begin{array}{c}
\text{OCH}_3 \\
\text{H}_3\text{CO}-\!\!\!\bigcirc\!\!\!-\text{OCH}_3 \\
| \\
\text{CH}_2 \\
| \\
\text{CH}_2 \\
| \\
\text{NH}_2
\end{array}
$$

Mescaline

it can produce a "model psychosis" in man and the symptoms may be altered or prevented by pretreatment with certain ataraxic agents (tranquillizers)(2). It develops no tolerance, and does not produce abstinence or withdrawal syndrome (3). It is used only for investigational and experimental purposes.

11.3 References

1. Emde, H. - Helv. Chim. Acta, 12:399 (1929).

2. Grollman, A. - Pharmacology and Therapeutics, 6th ed., p. 252, Lea and Febiger, Philadelphia, 1965.

3. La Barre, W., et al. - Science, 114:582-583 (1951).

4. Reti, L. - in R. H. F. Manske and H. L. Holmes (eds.): The Alkaloids, Vol. III, (a) pp. 324-329; (b) pp. 339-362, Academic Press, New York, 1953.

5. Welsh, L. H. - J. Am. Chem. Soc., 69:128 (1947).

6. Wilson, C. O., & Gisvold, O. - Textbook of Organic Medicinal and Pharmaceutical Chemistry, 4th ed., p. 394, Lippincott, Philadelphia-Montreal, 1962.

7. Wilson, C. O., & Jones, T. E. - American Drug Index 1964, pp. 262-263, Lippincott, Philadelphia-Montreal, 1964.

CHAPTER 12

ALKALOIDAL AMINES OF COLCHICUM

12.1 Colchicine

(a) Plant Sources

The corm, seed, and flower of the autumn crocus, Colchicum autumnale L., Fam. Liliaceae (also known as meadow saffron), contain a number of neutral, phenolic, and basic alkaloids as well as some fat-like substances. The best known of the alkaloids of Colchicum is colchicine, which occurs in the corm and the seed to the extent of 0.05 - 0.6%. The Colchicum Corm of the B.P.C. is specified to contain not less than 0.25% alkaloids.

Colchicine has also been found to occur in the corm of a number of other species, including Colchicum vernum Ker-Gawl., Gloriosa superba L., Merendera attica Boiss. et Sprun., Merendera sobolifera C.A.M. and Androcymbium gramineum McBr. (8).

(b) Properties and Reactions of Colchicine and Derivatives

Colchicine contains a tropolone part in ring C of its molecule. A number of chemical properties of colchicine are common to a number of other tropolones. Thus, ring C becomes benzenoid in the presence of base. Of the four methoxyl groups in the colchicine molecule, the one at C_{10} is much more easily hydrolyzed than the other three which are attached to the aromatic ring A (1, 8).

Colchicine is a very weak base; it is neutral to litmus. It may be extracted from both acid and alkaline solutions by shaking with chloroform. It is soluble in water, aqueous alcohol (less so in absolute alcohol), or

chloroform, and in hot benzene or hot amyl alcohol, but insoluble in ether
or petroleum ether (2). It may be crystallized from ethyl acetate. Its
optical activity is from the asymmetry of C_7. Colchicine and its derivatives show ultraviolet absorption at the 350 mμ. region.

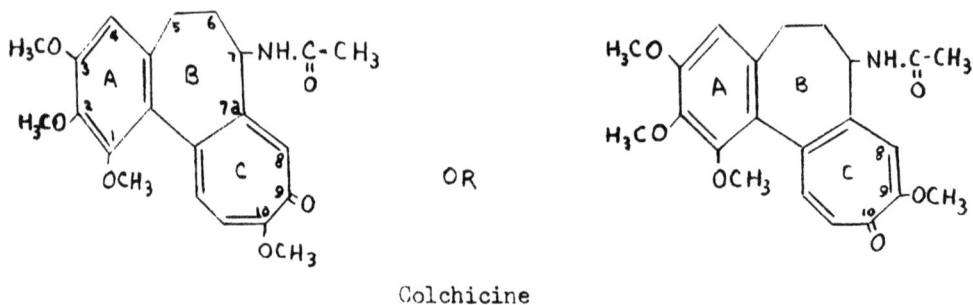

Colchicine

Mild hydrolysis of colchicine by aqueous hydrochloric acid solution
or dilute alkali gives methyl alcohol (from the methoxyl group of ring C)
and colchiceine. Colchiceine on more vigorous hydrolysis with hydrochloric
acid (d. 1.15) at 150° for six hours gives one molecule of acetic acid per
molecule of colchiceine and trimethyl colchicinic acid (1). The acetic acid
is derived from the side chain at C_7. The trimethyl colchicinic acid is an
amphoteric substance (with NH_2 group at C_7) and forms salt with both acid
and alkali. Methylation of colchiceine with diazomethane gives colchicine
and its isomer isocolchicine.

Tropolones generally form chelate complexes with many metals. So
does colchiceine which forms a crystalline copper salt, but not colchicine
(1, 8).

The methoxyl groups of colchicine may be replaced by amino, mercapto, or larger alkoxy groups through the actions of amines, mercaptans, and alcohols respectively. With guanidine or thiourea, colchicine produces imidazole derivatives.

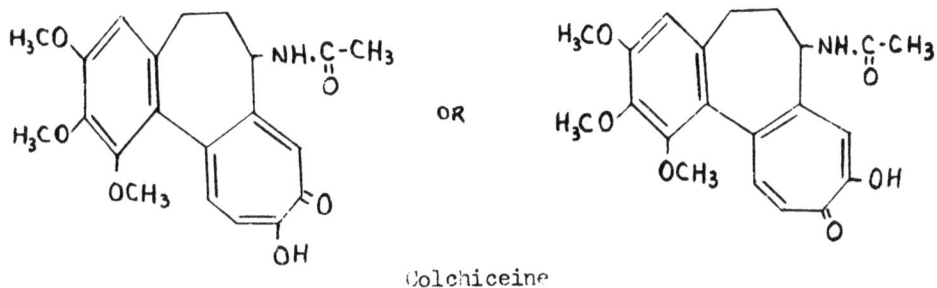

OR

Colchiceine

Colchicine's ring C becomes benzenoid in the presence of strong base. Thus, colchicine or isocolchicine in methanolic sodium methoxide gives allocolchicine (also called colchicic acid methyl ester).

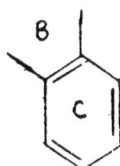

Allocolchicine
(Colchicic acid methyl ester)

In alkaline solution, bromination of colchiceine gives N-acetyldibromocolchinol. In acetic acid, bromination of colchiceine gives tribromocolchiceinic acid, which is decarboxylated in alkaline solution (8).

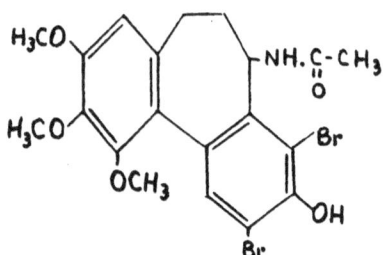

N-Acetyldibromocolchinol

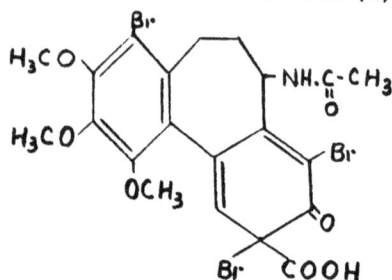

Tribromocolchiceinic acid

Colchiceine gives a green colour with ferric chloride. This has served as the basis of a colorimetric method for the quantitative estimation of colchicine (hydrolyzed to colchiceine) (6), but for this purpose the colchiceine must first be obtained in a purified form. There is a gravimetric method for quantitative estimation (see below). Colchicine and other tropolone ethers on developed paper chromatogram may also be located by first hydrolyzing them to colchiceine by spraying with 10% hydrochloric acid and heating the paper at 110° for two minutes and subsequently spraying with ferric chloride which gives green spots for the tropolones (1). These alkaloids containing the tropolone ring may also be detected on paper chromatograms by their fluorescence under ultraviolet light or by spraying with phosphotungstic acid reagent (7).

One system, among many others, of paper chromatography used for the separation of the Colchicum alkaloids is by impregnating Whatman No. 1 paper with 30% formamide in methanol or ethanol, and developing the chromatogram with benzene:chloroform (2:1) (4).

(c) Quantitative Estimation of Colchicine in Plant Material

As mentioned above, a colorimetric method (6) has been devised for estimation of colchicine on the basis of the colour reaction with ferric

chloride. The gravimetric method of Davies (2, 3) is claimed by some auth-

ors to be very accurate. The procedure may be summarized in outline shown

in Figure 9.

The phosphotungstic acid reagent used in this procedure consists of

10 gm. of sodium tungstate, 6 gm. of sodium phosphate, and 50 ml. of water

acidified with nitric acid.

(d) Isolation of Colchicine

The isolation procedure (8) may be summarized in outline as shown in

Figure 10.

(e) Pharmacological Action and Uses

Colchicine, administered orally or intravenously, relieves the pain

and inflammation around the joints in an acute attack of gout. It is said

to be effective in over 90% of patients. The mechanism for this action re-

mains obscure. It appears to exert no beneficial effect in the intervals

between acute attacks (5).

Colchicine arrests cell mitosis at the metaphase, preventing the

formation of the spindles. This antimitotic action is utilized in certain

studies in genetics and in induction of polyploidy in plants. Colchicine's

effectiveness in gout apparently is not due to its antimitotic action. The

colchicine derivative trimethylcolchicinic acid has no antimitotic action

but is effective in the treatment of acute gouty arthritis. Colchicine has

also been used for its antipyretic action in Hodgkin's disease. Larger

doses of colchicine may cause diarrhea and pain in the gastric region.

Colchiceine (demethylated product of colchicine, by mild acid hydro-

lysis) is ineffective in gout but has been used in leukemia. N-desacetyl-N-

Powdered plant material (5 gm.)

↓

Percolation with 100 ml. of 70% alcohol

↓

Remove solvent by evaporation

↓

Residue taken up with water

↓

Shaken with several portions of chloroform

↓

Remove solvent

↓

Dissolve residue in hot water (30 ml.)

↓

Filter

↓

To cooled filtrate, add 3 ml. phosphotungstic
acid reagent and 7 ml. dilute sulphuric acid

↓

Centrifuge

↙   ↘

Supernatant   Precipitate

↓

Suspend in water

↓

Shake with chloroform
(30 ml. portions)

↓

Remove chloroform by evaporation
with the help of alcohol

↓

Dry residue of colchicine to
constant weight and weigh

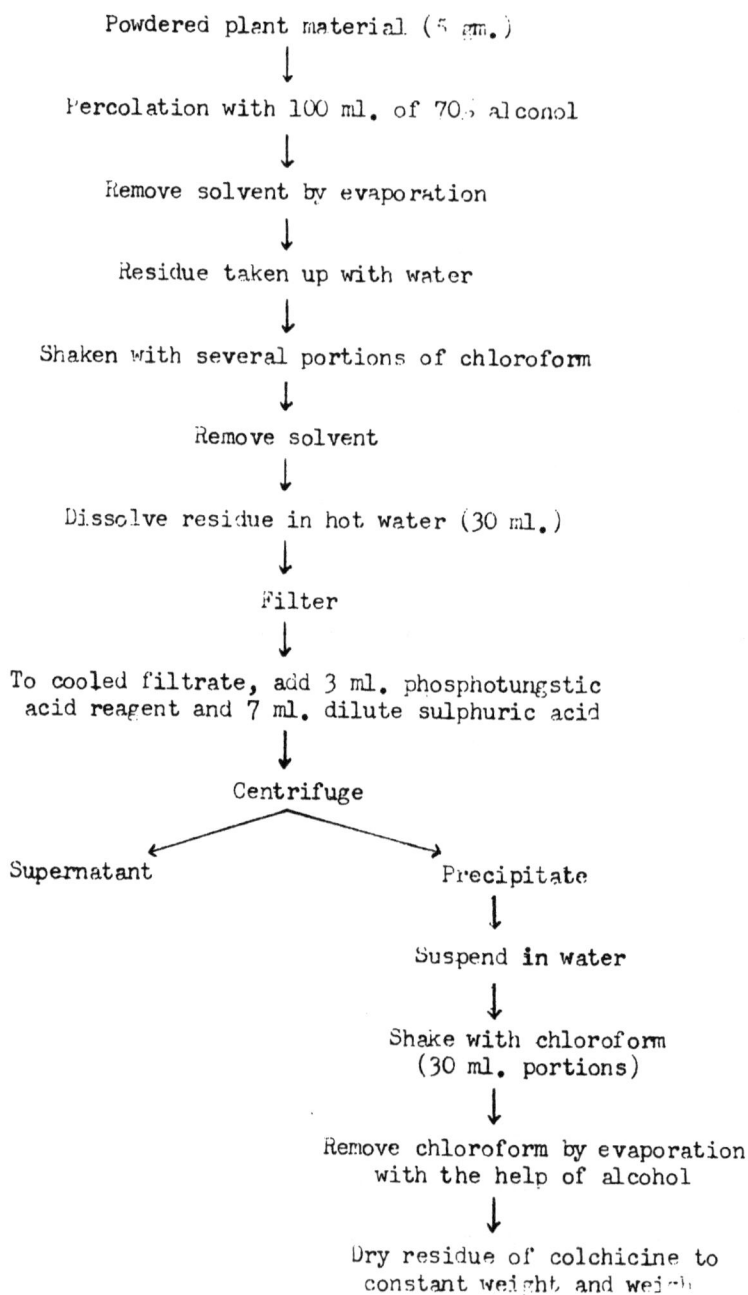

Figure 6.   Procedure Scheme for Quantitative Estimation of Colchicine

- 174 -

Powdered Corm (20 Kg.)
↓
A series of extractions with a total of
60 litres of ethanol
↓
Concentrate under reduced pressure
↓
Dilute with water to 4 litres
↓
Wash with several 2-litre portions of ether
↓
Acidify the aqueous solutions with HCl to pH 2 - 3
↓
Repeated extractions with 500-ml. portions of chloroform
↓
Remove solvent from combined chloroform extract
↓
Residue (48 gm.)
(neutral and phenolic chloroform-soluble material)
↓
Dissolve in minimum amount of ethanol
↓
Treat with 500 ml. water and add 10 gm. NaCl
↓
Filter

Brown ppt. ←   → Filtrate
(Apigenin)

Extract with five   Chromatographed on
200 ml. portions of   600 gm. of alumina
chloroform   ↓

Aqueous phase   Chloroform   Elute with ether:
↓   extract   chloroform (2:1
Dry residue   and 1:1)
↓   ↓
Extract with chloroform   Crude colchicine
↓   (16.54 gm. from 37
Remove solvent   gm. of neutral and
↓   phenolic CHCl$_3$-sol.
Residue (colchiceine)   material
↓   ↓
Crystallized from ethyl acetate   Re-crystallized
                                     from ethyl acetate
                                     and ether to give
                                     14 gm. colchicine

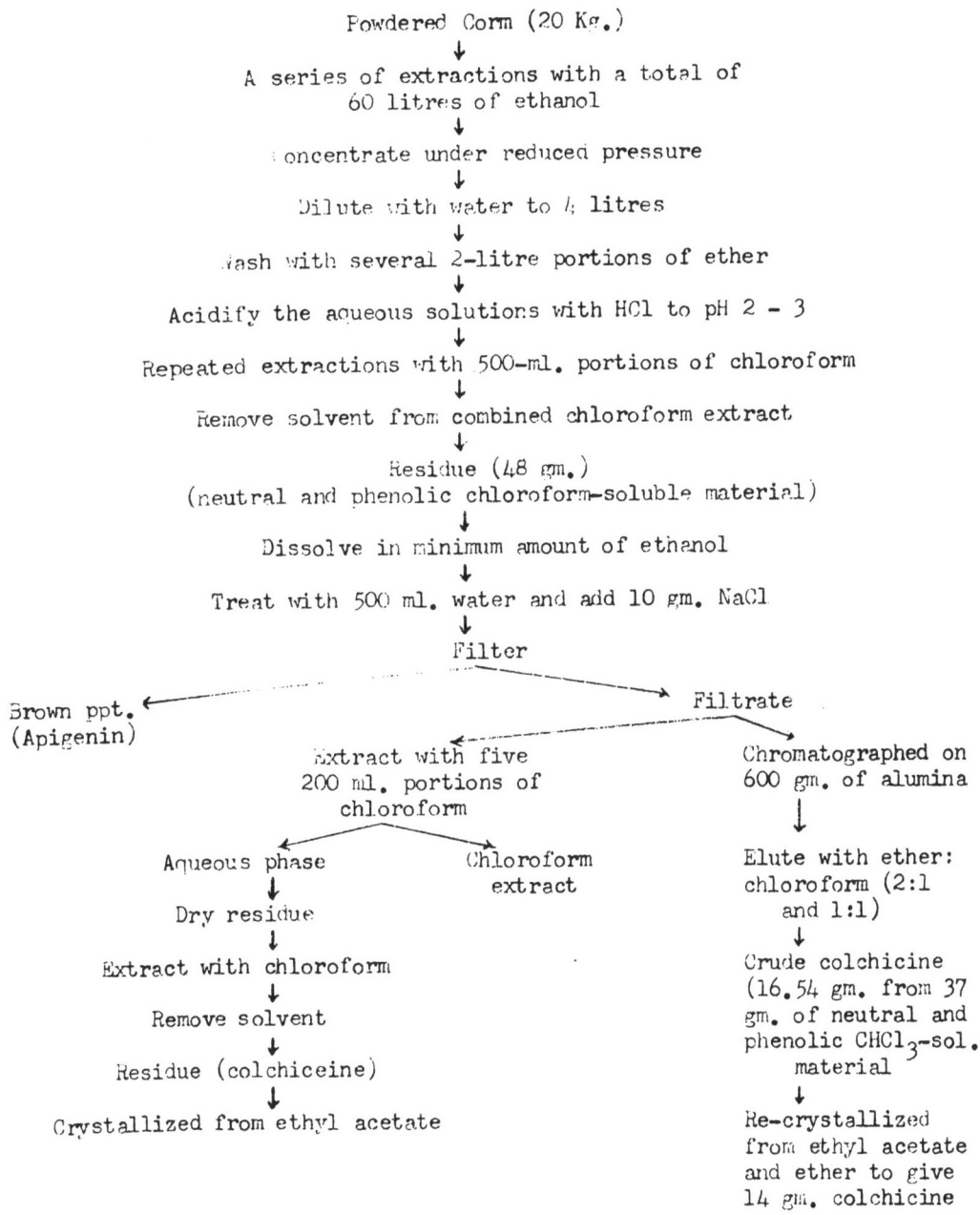

Figure 10. Procedure Scheme for the Isolation of Colchicine

methylcolchicine (also known as demecolcine), which is a basic alkaloid occurring as such in many species of Colchicum, Gloriosa, and Merendera (8) shows a depressant effect on granulocytes and has been used in the treatment of chronic myeloid leukemia (5).

(f) Medicinal Preparations

        Colchicum Corm B.P.

        Colchicum Tincture B.P.

        Colchicum Liquid Extract B.P.

        Colchicine U.S.P. XVII, B.P. 1963

12.2 References

1. Cook, J. W., & Loudon, J. D. - in R. H. F. Manske and H. L. Holmes (eds.) The Alkaloids, Vol. II, pp. 261-329, Academic Press, New York, 1952.

2. Cromwell, B. T. - in K. Paech & M. V. Tracey (eds.): Modern Methods of Plant Analysis, Vol. IV, pp. 432-436, Springer, Berlin, 1955.

3. Davies, E. C. - Pharm. J., 106:480 (1921).

4. Delong, V., Havrlikova, J., & Santavy, F. - Ann. Pharm. Franc., 13:449-454 (1955).

5. Grollman, A. - Pharmacology and Therapeutics, 6th ed., pp. 178-180, Lea and Febiger, Philadelphia, 1965.

6. King, J. S. - J. Am. Pharm. Assoc., Sci. Ed., 40:424-427 (1951).

7. Walaszek, E. J., Kelsey, F. E., & Geiling, E. M. K. - Science, 116:225-227 (1952).

8. Wildman, W. C. - in R. H. F. Manske (ed.): The Alkaloids, Vol. VI, pp. 247-283, Academic Press, New York, 1960.

Festuclavine, 104-106

properties of, 26-27
Norlupinane, see Quinolizidine
Noscapine, 57-60, 62, 64
Nux Vomica, 124, 132

O

Opium alkaloids, 54-65
    action and uses of, 63-65
    biosynthesis of, 61, 63
    medicinal preparations of, 65-66
    separation of, 61, 62
    tests for, 56, 58

P

Papaver somniferum, 54
Papaverine, 58-65
    action and uses of, 64-65
    biosynthesis of, 61, 63
    properties of, 58-59
    tests for, 58, 59
Paspalum distichum, 89
Penniclavine, 105-106
Pennisetum typhoideum, 103
Peyote, 167
Phenethylamine alkaloids, 163
Physostigma venenosum, 132
Physostigmine, 132-134
Pilocarpine, 159-161
Pilocarpus jaborandi, 159
Pilocarpus microphyllus, 159
Pilocarpus pennatifolius, 159
Pilocarpus racemosus, 159
Protoveratrines A and B, 151-152,
    155-157
Protoverine, 151-152
Pseudoephedrine, 163, 164, 166
Pseudojervine, 150
Pyridine alkaloids, 25
Pyroclavine, 104-105

Q

Quinidine, 75, 78-82
    action and uses of, 82
    isolation of, 80, 81
    properties of, 78-79
Quinine, 75-82
    action and uses of, 80
    isolation of, 80, 81
    properties and tests, 76-79
Quinoline alkaloids, 75
Quinolizidine, 83, 85

R

Rauwolfia alkaloids, 113-124
    action and uses of, 122
    grouping of, 116
    isolation of, 121-123
    medicinal preparations of, 124
    properties of, 116-120
Rauwolfia canescens, 114, 121
Rauwolfia densiflora, 115
Rauwolfia heterophylla, 114
Rauwolfia hirsuta, 114
Rauwolfia micrantha, 115
Rauwolfia nitida, 115
Rauwolfia perakensis, 115
Rauwolfia schueli, 115
Rauwolfia sellowii, 115
Rauwolfia serpentina, 113-116,
    120-122, 124
Rauwolfia tetraphylla, 114-115,
    121
Rauwolfia vomitoria, 115-116
Recanescine, see Deserpidine
Remijia pedunculata, 75
Remijia purdieana, 75
Reserpine, 114-124
    action and uses of, 122, 124
    isolation of, 120-123
    properties of, 116-120
Rescinnamine, 114-124
Reticuline, 59, 60, 63
Rubijervine, 149

www.ingramcontent.com/pod-product-compliance
Lightning Source LLC
Chambersburg PA
CBHW051754200326
41597CB00025B/4549